THE DRAGON'S TAIL

THE DRAGON'S TAIL

Radiation Safety in the
Manhattan Project, 1942–1946

BARTON C. HACKER

UNIVERSITY OF CALIFORNIA PRESS
Berkeley   Los Angeles   London

University of California Press
Berkeley and Los Angeles, California

University of California Press, Ltd.
London, England

Library of Congress Cataloging-in-Publication Data

Hacker, Barton C., 1935–
  The dragon's tail.

  Bibliography: p.
  Includes index.
  1. Nuclear weapons—Safety measures. 2. Ionizing radiation—Safety measures. 3. United States. Army. Corps of Engineers. Manhattan District—History.
  I. Title.
U264.H33   1987   363.1'79   86-11427
ISBN 0-520-05852-6 (alk. paper)

Printed in the United States of America

1 2 3 4 5 6 7 8 9

CONTENTS

Foreword

The Department of Energy's Nevada Operations Office (DOE/NV) conceived this research project in the fall of 1977. At that time many questions and concerns were being raised by the public and Congress regarding nuclear test participants' radiation exposures and radiation safety practices in the nuclear weapons atmospheric testing program. No single document describing radiation safety practices in the nuclear test program then existed, but the need to develop such a document was becoming obvious. Hence we initiated the development of a manuscript that would (1) document the development of radiological safety in the nuclear weapons testing program, (2) be easy enough for the layman to understand yet contain enough specific information to make it useful for technical purposes, and (3) capture the reader's interest.

Time was of the essence; oral interviews had to be conducted with key former participants who were aging—many already in their seventies. Written documentation existed but was spread out over various locations across the country, and many documents were approaching the end of the required document retention period. Furthermore, a historian knowledgeable in radiation sciences proved rare; the person hired for the job had to undergo a steep learning curve. Compounding the time constraints, the historian had to be investigated for a DOE security clearance to allow him access to classified documents. In the spring of 1978 the author/historian was hired by Reynolds Electrical & Engineering Co., Inc., a DOE/NV prime contractor, to develop this comprehensive history. With the administrative details taken care of, Dr. Barton C. Hacker's efforts were under way by mid-1978.

As the early chapters were developed, it became evident that the initial two-year completion goal could not be met if the manuscript was to fulfill our original objectives. Integration of oral interview information with written documentation, all properly referenced, was a more complex task than had been expected. The chapters, once written, had to be reviewed by DOE/NV staff and a peer review group consisting of some fifty prominent scientists and key test participants. Once all the comments had been received, Dr. Hacker then had to resolve any conflicts. The entire process consumed much time but was vital if the product was to be the most accurate historical account possible.

By 1984, although ten of the fifteen chapters were at some stage of completion, it was evident that at least some of the completed chapters had to be made available to the public soon. Since the late 1970s many legal claims had been placed against the government in connection with the nuclear weapons atmospheric testing program. Information contained in this manuscript could provide both plaintiff and defense counsels an accurate perspective on radiological safety practices in the early testing years. It was decided that the conclusion of the Manhattan Project would make a good cutoff point, and efforts proceeded to complete the final review process of the history of radiation safety from 1942 through 1946. That process consisted of a final technical review on December 5 and 6, 1984, by a select panel of scientists and historians.

A special note of acknowledgment and appreciation is extended to those individuals who have given of their time to be interviewed and to those who reviewed the manuscript. The continued support from the management of the Nevada Operations Office and DOE Headquarters program officials has been vital to this task and is acknowledged. The program supervision extended by members of the Health Physics Division staff has been critical and the efforts of Messrs. Michael A. Marelli and Marshall Page, Jr., are specially noted. The extensive effort of Dr. Barton C. Hacker is also acknowledged. Without the dedication he has displayed, it is doubtful that such work as this would have been possible. It is my hope that the development of both manuscripts will provide the information to remove much of the speculation and wonderment often expressed about the radiological safety program for the testing of nuclear weapons.

Bruce W. Church, Director

Health Physics Division
Nevada Operations Office
Las Vegas, Nevada

Preface

This volume benefits from the institutional and financial support of the United States Department of Energy, successor to the Atomic Energy Commission and the Energy Research and Development Administration. In 1977 officials from DOE's Nevada Operations Office conceived the need for a complete and reliable historical account of radiological safety in nuclear weapons testing. Reynolds Electrical & Engineering Co., Inc. (REECo), the support firm that has operated the Nevada Test Site for the Atomic Energy Commission and its successors for more than three decades, hired me in 1978 to write that history. It has grown into a far larger project than any of us anticipated, this volume having become but the first of two. The second, which is well under way, traces the history of radiation safety in testing nuclear weapons from the formation of the AEC after World War II until the present.

I have received much guidance from participants in the events I describe, not only in Nevada but everywhere in the country. Many individuals have allowed me to interview them, have read and commented on all or part of the manuscript, and have encouraged or otherwise aided me in this work. The names of those I have interviewed, many of whom also commented on draft chapters, appear in the list of interviews at the back of the book, but I here wish to extend my heartfelt thanks for their help.

For reading the manuscript and otherwise contributing to the endeavor, the following members of REECo's Environmental Sciences Department deserve acknowledgment: Arden E. Bicker, Department Manager; B. Lee Brown, former Dosimetry Research Project Manager; Bernard F. Eubank, Technical Information Officer; Linda K. Jensen, Dosimetry Research Project Staff Assistant; Linda M. Rakow, Dosimetry Research Project Manager; Roger C. Thompson, former Laboratory Analysis Superintendent; and Ira. J. Wells, Senior Health Physicist.

Special thanks must go to William J. Brady, Department Technical Advisor, who shared with me his knowledge of nuclear weapons testing and critically reviewed every one of my drafts; he saved me from more mistakes than I care to contemplate.

Mr. Brady also chaired the review panel, all of whose members I wish to thank for their painstaking and detailed examination of the final draft of *The Dragon's Tail* for accuracy, integrity, and readability. Other members of the review panel were Robley D. Evans, Professor Emeritus of Physics, Massachusetts Institute of Technology; Louis H. Hempelmann, Jr., M.D., Professor of Radiology, University of Rochester, and former Health Group Leader, Los Alamos National Laboratory; Richard G. Hewlett, History Associates Incorporated, and former AEC and DOE Chief Historian; John S. Malik, Scientific Advisor for underground testing, Los Alamos National Laboratory, and Test Panel member, Nevada Test Site; and William H. McNeill, Robert A. Millikin Distinguished Service Professor of History, University of Chicago.

Many other current and former members of DOE, the Defense Nuclear Agency, and their contractors have also assisted me in various ways. So, too, have the many scholars and interested laypersons from whose comments and other help I have further profited. All are included in the list of reviewers. I wish to thank them, as well as those named in the list of interviews.

Last and most important, I thank Sally L. Hacker, although thanks seems too inadequate a word. Professional responsibilities, hers and mine alike, have separated us physically, but distance has never diminished her support—emotional, spiritual, and intellectual—for a project that more than once might otherwise have seemed discouragingly endless and thankless.

Whatever virtues this book may attain owes much to those who have given so freely of their time and knowledge. Its flaws reflect, in part, my difficulty in taking full advantage of all the advice I have received. In the final analysis, I bear full responsibility for what I have written, but no one knows better than I that it could not have been done unaided.

Las Vegas, 1984   B. C. H.

Introduction

RADIATION SAFETY IN WORLD WAR II

The central Pacific morning of 25 July 1946 dawned bright and clearing over Bikini atoll. Scientists finished checking gear aboard barge LSM-60 and left the lagoon just after six o'clock. Ninety feet below the barge hung a watertight caisson holding an atomic bomb; it was the fifth such device ever used, all in little more than a year, but the last for many months to come. Above the barge towered a radio antenna to receive the coded firing signal at How hour, 8:35 A.M., two and a half hours later. As the moment neared, only a few low-lying clouds and scattered high cumulus lingered as reminders of predawn rain squalls and lightning. On the lagoon's calm surface floated a mighty fleet, seventy-four vessels ranging from aircraft carrier and battleship to concrete barge and landing craft. The vessels carried stores, munitions, and fuel, but no voice broke the stillness, no human being stirred. It was called the ghost fleet: an anchored armada arrayed to show the effects of a shallow underwater atomic explosion. A cautious distance away at sea, Joint Army-Navy Task Force One awaited Baker, the second test of Operation Crossroads.

Caution proved well advised. The bomb performed as expected, not surprising for the same plutonium-fission model already tried near Alamogordo and over Nagasaki the year before, and in Crossroads Able earlier that July. The surprise came in what followed. As thousands upon thousands of tons of water collapsed back into the lagoon, a surging wall of radioactive mist blanketed the ghost fleet. "These contaminated ships became radioactive stoves," observed the military committee assigned to evaluate the test. Dismayed salvage teams could scarcely approach most of the target vessels for days; some ships remained off limits much longer. Worse than aborted salvage plans, spreading contamination began to threaten the task force's own support ships and crews. Vigorous efforts by the radiological safety section forestalled apparent harm to any member of the task force, but many safety experts saw too close a call for comfort.

Bombs bursting in air—like the one over Bikini lagoon in the first Crossroads test—caused few radiological safety problems. Radioactive debris (little enough in any case) rose too high, stayed aloft too long, and scattered too widely to pose any danger when falling finally to earth. These results may have helped obscure the warning implicit in the first test of an atomic bomb, Trinity, in July 1945, a warning already muted by good luck and wartime secrecy. Detonated atop a hundred-foot tower and sucking huge amounts of earth and other debris into the rising fireball, it produced enough fallout to alarm those in charge. Superficially burned livestock, however, suffered the only certain damage. As the danger passed, urgent wartime demands discouraged dwelling on might-have-beens. Crossroads Baker left no such easy exit. Testing nuclear weapons clearly posed hazards more severe than even the most expert scientists had fully anticipated.

Radiation safety had been recognized fifty years earlier as a far more modest problem. Before World War II, those chiefly at risk were limited numbers of doctors and technicians working with X-ray machines or radium. Safety centered on the concept of tolerance, the amount of radiation living systems could absorb without irreparable damage. Tolerance doses measured in roentgens became the formal basis for self-imposed X-ray safety codes by the mid-1930s. Concern aroused by the fate of dial painters and other victims led to such codes being extended to radium, then to radon. Standards for tolerable body burdens of radium and for air concentrations of active gases were published in 1941. These three basic standards—for X rays and gamma rays, for ingested radium, and for airborne radon—framed the structure of safety in the fission-bomb project. But much remained unknown or uncertain as the United States launched what the postwar world learned to call the Manhattan Project, the bold venture to design and produce the first atomic bombs.

When the project began in 1941, some basic properties even of the natural radioelements remained question marks; their effects on living things were still less clear. Yet they were well known compared with properties of the active materials used in the bomb project—plutonium, enriched uranium, the products of fission—materials that would require handling in amounts vastly larger than all the radium ever used. Special concerns centered on making plutonium and culling it from fission products. Piles that converted uranium to plutonium produced radiation levels far higher than any met in the past. Plutonium itself, discovered only in 1941, was a complete unknown. And danger might now threaten not merely doctors and technicians by the hundreds, but workers of all kinds by the tens of thousands. Most would know little about the dangers, and few could be told much in a highly secret project. Possible risks to the public compounded the problem, as larger numbers of potential victims offset the presumably lesser danger to any single person.

The bomb makers foresaw these problems. Knowing the hazards, they accepted the need to protect workers and the public against undue exposure. Soon after starting work on plutonium, therefore, the Metallurgical Laboratory of the University of Chicago formed a health division as full partner to its Physics, Chemistry, and Engineering divisions. The Health Division first adapted existing standards and practices to safeguard Chicago project workers. It also launched research programs to learn more about the hazards and the means of coping with them. Ultimately, the entire Manhattan Project modeled its health and safety effort on Chicago practice and research. Chapter 1 surveys the base upon which the Health Division built, the half century of attempts before World War II to counter the hazards of X rays and radium and to devise workable standards for their safe use.

Health Division experts believed they could hold radiation risks, the "special hazards" in project usage, to tolerable levels. By and large they did, but whether they did or not might have mattered little. Health hazards did not rank highest among the risks of gambling large amounts of money, men, and scarce resources on the effort to convert untested theories into working weapons. Early in World War II, no risk outweighed the threat of a Nazi bomb. A German laboratory was the scene of the first recognized nuclear fission in 1939. After the war began, Allied intelligence believed Germany was backing further research. The prospect of seeing Nazi Germany first to make so awesome a weapon was a horror against which almost any risk seemed worth taking.

Health and safety, in fact, may well have posed some of the least baffling problems to confront the Manhattan Project. Routine industrial safety proved to require the more intensive effort, though deeper concern centered on the special hazards. In the event, prewar standards broadly sufficed, despite the host of new and exotic active substances produced in the piles. Although applied on a vastly larger scale, the trial-and-error methods devised over four decades to protect radiation workers mostly met the need. Radiation sources were shielded, careful work habits instilled, workplaces closely watched, workers screened to detect early signs of damage. Setting standards and seeing them enforced were the first tasks of the Health Division.

Those responsible for project health and safety had other concerns as well. Much Health Division research addressed urgent questions about handling and using the many unfamiliar substances the project required. Researchers also explored problems more likely to matter in the longer run. How valid were extant standards? Were all the risks, toxicological as well as radiological, fully perceived? Did new and strange substances pose unknown risks? Were there better ways to detect radiation? Could radiation-caused damage be treated if not prevented? Such questions evoked a wide-ranging research program that sometimes threatened to overshadow more routine work in clinical medicine, health physics, and industrial safety. Health Division opinion was itself divided about the wisdom of spending scarce resources on basic research while war raged; army officers running a crash program showed even less enthusiasm. Although research produced results—better instruments, enhanced knowledge, even new protection concepts—safety remained closely modeled on prewar theory and practice. The main difference was the vastly larger scale of the wartime work, which meant vastly enlarged hazards. Chapter 2 recounts Chicago Health Division efforts to balance wartime demands against long-term safety.

Throughout the Manhattan Project, radiological safety began with good housekeeping. Careful handling of active subtances and proper shielding against radiation largely protected workers from dangerously high levels of contamination or exposure. But turning fissile metals into bombs posed new, and sometimes greater, dangers. The people assigned to the weapons laboratory at Los Alamos in northern New Mexico first faced the unprecedented hazards of making a nuclear bomb. Starting from scratch under intense wartime pressures, they struggled to devise techniques for treating, purifying, and shaping plutonium, a metal neither they nor anyone else knew very much about. What dangers they might face, what safeguards they needed, were matters of guesswork, informed only by extant radium experience and standards. Protecting laboratory workers became the first task of the small Los Alamos Health Group.

Laboratory crises, however, paled before other strains on Health Group resources when the first plutonium-bomb design proved unworkable in mid-1944. Developing new designs entailed a series of hazardous open-air experiments in the canyons around Los Alamos during the next six months. A young scientist likened one of those experiments to "tickling the tail of a sleeping dragon." Probably no more than a casual quip, the comment nonetheless tapped a deep vein of meaning. In the lore of dragon hunting, the reptile's potent tail posed a special danger. Its thrashing might inflict unexpected damage on the unwary hero even after the animal itself had received its death wound. Metaphorically, the dragon's tail aptly symbolizes the "special hazards" of the Manhattan Project and, perhaps even more aptly, the key process of testing nuclear weapons.

The experimental findings in late 1944 and early 1945 resolved some doubts, but not all. Bomb makers felt compelled to test their gadget. "Gadget" was the word they used. One reason, of course, was secrecy. Perhaps so homely a word also eased the minds of men and women facing a monstrous unknown. In any event, testing the first bomb posed safety problems far more serious than even the broadest meaning of good housekeeping might span. Planning had been under way for a year when the decision for a full-scale field test, code-named Trinity, was made in March 1945. Although the desert site in southern New Mexico was chosen, in part, for the sparsity of its nearby population, the most violent man-made explosion in history would still pose public danger. Radioactivity might pose even more serious hazards, and workers at the test site might not be the only ones at risk. Clouds of plutonium dust or fission products could threaten health and lives miles downwind from the blast. When they decided to conduct the test, planners judged other factors to outweigh the potential dangers. Minimizing the risks fell to the Health Group. Chapter 3 describes how the group planned to do so, and how it coped with the other challenges that carried Los Alamos to the brink of Trinity.

Initially at least, testing the gadget seemed a difficult but straightforward task. Priority inevitably centered on making certain the bomb worked. Security ran a close second. Proving the bomb while keeping it secret largely dictated the shape of Trinity plans. Through the spring of 1945, however, neither science nor secrecy seriously conflicted with safety. Everyone assumed that safety meant chiefly protecting test workers, relatively few in number, subject to close control, and knowing the hazards; they could be expected to obey rules designed to safeguard their own well-being. The only likely public threat was bad weather or winds blowing the wrong way. Scheduling based on long-range forecasts for perfect weather and winds would avert that danger. Then, in mid-June, these assumptions collapsed. As closer study raised new questions about public safety, wartime diplomacy impinged on the timing of the shot.

If worst came to worst, new calculations suggested, fallout beyond the test site might be far more severe and widespread than anyone had supposed. Civilians, perhaps hundreds of them, might require evacuation; public safety, in other words, might threaten security. Worse yet, safety plans could no longer rely on perfect weather. President Truman expected to issue an ultimatum to Japan from his Potsdam meeting with other Allied leaders scheduled to begin 16 July. Germany had surrendered in May, but neither the president nor any of his advisors hesitated to wield the yet unproved weapon against the surviving Axis partner. Los Alamos shipped the only uranium bomb, code-named Little Boy, to the Pacific early in July. Straightforward in design, this model needed no test, but it did require large amounts of enriched uranium, still in short supply. Another bomb could not be ready for months. Credibility demanded something quicker; that meant Fat Man, the first plutonium bomb, but Trinity must first prove it would work. Contrary to early plans, doubtful weather would no longer automatically postpone the test. Contingency planning suddenly became urgent as the once smooth ties linking science, security, and safety frayed.

External pressures, for the most part, scarcely troubled work at Trinity. Preoccupied with mounting the test, workers wasted little time thinking about politics, diplomacy, or the conduct of the war. Bureaucratic and technological momentum played far larger roles in bringing the project to its climax. Bad weather on the eve of the test delayed firing only briefly. At 5:30 in the morning of 16 July 1945, the radiance of the first atomic bomb outshone the dawn. The implosion bomb proved to be all its makers had hoped, and more. The glow of accomplishment ended a job well done for the physicists. For the Health Group it signaled the start of a grueling and sometimes alarming day. Radiation monitors found much what they had expected on the test site itself. Not so beyond Trinity's borders, where trouble loomed in the form of disquietingly high levels of local fallout. Hours elapsed before monitors were able to convince themselves that fallout everywhere remained within safe limits. Eventually the Health Group decided that no one had suffered any lasting harm from the test. Like other aspects of Trinity planning, safety met the needs of wartime testing. Chapter 4 tells how.

Just three weeks after Trinity burst over the New Mexico desert, another mushroom grew over Hiroshima. Nagasaki met the same fate three days later. Within a month of Trinity, World War II ended, in no small measure the result of what Trinity proved. Tens of thousands perished in the two cities: the effects of blast, heat, and ionizing radiation killed many outright; others succumbed more slowly to wounds. Japanese rescue and relief efforts began at once, bolstered with American aid when occupation forces arrived. American teams also joined Japanese to study the bombings and the victims. Catastrophe, as sometimes happens, produced a wealth of scientific data otherwise beyond reach. Like the unfortunate radium dial painters before World War II, the victims of Hiroshima and Nagasaki taught us much about long-term human costs of radiation.

In the aftermath of the Japanese bombings, however, military planners found long-term effects of small concern. Far more pressing seemed the problem of atomic bombs and the conduct of future war. Enter Operation Crossroads. Domestic politics and international relations in a troubled postwar era no doubt shaped the decision, but testing at Bikini atoll in the central Pacific began with hard questions about tactics, troops, and machines in nuclear war. A joint army-navy task force conducted Crossroads, a major operation mounted 4,500 miles from American shores: Joint Task Force One comprised 42,000 men and women, 242 ships, 156 aircraft. Radiological safety in so huge an undertaking at so remote a site posed massive problems. Acting for the army-navy task force, the Manhattan Engineer District assumed responsibility, naming the longtime chief of its own medical office to head the radiological safety section. How the section planned to apply the lessons of Trinity and Japan at Bikini is the subject of chapter 5.

In July 1946 two atomic bombs, the same kind used the year before on Nagasaki, were fired at Bikini: one dropped from a B-29 to burst hundreds of feet above the target fleet; the second, suspended from a barge, exploded beneath the surface of Bikini lagoon. Test Able produced no startling surprises, but the aftermath of Baker later that month stretched equipment, men, and technique to the limit. Radioactive water inundated the target fleet, leaving some ships unapproachable for weeks. Persistent radioactivity hampered salvage crews trying to restore contaminated target vessels to working order, and their efforts achieved only spotty results. A few target ships sailed from Bikini under their own power, but most of them had to be towed when the task force abandoned Bikini for nearby Kwajalein. Three-fourths of the target fleet never left the Marshall Islands at all, either sunk in the tests or destroyed afterward as unsalvageable. Only twenty-two of the original ninety-three target vessels eventually dropped anchor in west coast ports, where they underwent intense study.

Contamination after Test Baker also spread to the other vessels of Joint Task Force One. Radioactivity accumulated from seawater cycling through ship systems created one hazard, radioactivity concentrated by marine life on ships' hulls another. Strenuous efforts to protect work crews and other task force members from overexposure to radiation mainly succeeded, but the record fell far short of perfect. Indications of widespread contamination from unfissioned plutonium proved the final straw. Decontamination work ceased at the strong urging of the radiological safety section and the task force withdrew from Bikini in some haste. Eventually the navy launched a special program for clearing as safe every ship that spent any time in Bikini lagoon after Baker. Operation Crossroads posed more questions than it answered, particularly with respect to radiological safety. Chapter 6 describes what happened at Bikini and what it meant to those involved.

Manhattan Project experience transformed the context of radiation protection. Specialists in that field acquired a unique title, health physicist, and government controls largely replaced self-regulation. Acceptable limits for radiation exposure have since fallen lower than those permitted by the Manhattan Project. Risks from radiation have become more precisely defined, in both physical and human terms. Radiation can now be detected and measured more accurately; injuries can be treated more effectively. Yet the men who ran Manhattan Project safety programs clearly perceived both the dangers and the safeguards. Later research and new viewpoints have refined, rather than displaced, older theories. Ambiguity and uncertainty persist, but only partly because final answers still elude science; safety standards present social as well as technical problems. The epilogue reviews both the general problems of setting standards for radiation protection and the historical impact of the Manhattan Project on the process.

Radiation safety problems, of course, extended far beyond the boundaries of weapons testing. Possible sources of danger lurked at every stage in the process of making and storing nuclear weapons. Uranium miners, for instance, were known to risk higher rates of cancer from breathing radon underground. Disposing of nuclear wastes, at the end of the cycle, posed well-understood dangers. During World War II neither of these problems spurred as strong an effort to resolve as their importance warranted. I have little more than mentioned them. Nor have I systematically addressed the problems that arose in research and production phases of the Manhattan Project, although these became matters of far-reaching concern. Specific circumstances, in each instance, created unique problems, and telling the full story would draw me far afield. Throughout this work I have tried to say enough about these issues to make clear the context of test safety. Radiation safety in testing nuclear weapons remains the focus of this account.

1

FOUNDATIONS OF MANHATTAN PROJECT RADIATION SAFETY

ORIGINS OF PROTECTION STANDARDS

As 1895 ended, Wilhelm Conrad Röntgen, professor of physics at the University of Würzburg, published his discovery of X rays and explained how to produce them. He made headlines worldwide. Photographs like windows opened into living human bodies amazed the public, but science was no less excited to learn that the physical world still held real surprises. Above all, the new rays promised a tool of immense value in diagnosing and treating human disease. Within a year every major city in the United States had at least one medical X-ray center, and many machines appeared in private offices as well. Users had already reported side effects on people exposed to the rays. Most of them, such as reddened skin or lost hair, seemed minor. The worst resembled severe burns. Biological damage might not, however, be limited to burns. By 1903 doctors knew that X rays sometimes caused deeper effects as well: cancer, sterilization, damage to the blood-forming organs. Medical uses of X rays, clearly of great value, just as clearly entailed serious risks.[1]

The injured, chiefly X-ray workers themselves, deemed the price small for such large returns. But rising public concern, legal actions, and threatened legislative sanctions compelled practitioners to address the safety issue. Better machines and techniques improved the record. Increasingly more aware of the danger, X-ray workers also tried to limit unneeded exposure. As Daniel Paul Serwer has observed, the key was "common sense informed by experience."[2] The first decade saw proposed, if not widely adopted, almost every technique since used to prevent or control harmful effects of radiation. Then as now, however, the issue was not simply prudent caution to forestall clear danger. Although using X rays might be risky, greater safety could mean poorer results. Lacking precise measures either of hazards or gains, workers relied on their own judgments to strike some balance.[3] "To work without protection is foolhardy and inexcusable," remarked Dr. W. S. Laurence at a meeting of the American Roentgen Ray Society in 1907, "but I believe attempts at absolute protection have been carried to rather absurd extremes."[4]

Safety long remained a matter of personal choice and judgment. Trial and error, art more than science, governed the early use of X rays. The problem began with erratic machines. Although improved in detail, for nearly two decades after Röntgen X-ray tubes retained the same basic design: electrodes at either end of a sealed glass tube holding highly rarefied gas. The gas provided ions which the electrodes accelerated to produce X rays. In practice such tubes suffered two major drawbacks: gas pressure varied unpredictably, as did the voltage maintained across the electrodes by a static machine or induction coil. Each machine was unique, results from one hard to compare with those from another.[5] Worse yet, even using the same machine might not help. Pioneer radiologist Edith H. Quimby observed that "the output of radiation fluctuated from exposure to exposure, and even from minute to minute."[6]

Practitioners unsure how best to measure dose, or even what dose meant as applied to X rays, compounded the problem. Machine output, energy transmitted, and energy absorbed were all related to dose but were affected by diverse factors. Erythema (reddened skin) and epilation (hair loss) served as handy indexes for what X-ray workers really wanted to know: how much radiation tissues absorbed (biological dose in later terminology). Seldom expressed that way before the 1920s, the meaning of dose remained unclear for many years more. The concept of dose first entered X-ray work to protect patients. Found to vary in their effects on tissues, X rays could be applied to the treatment of a number of ailments. Cancer cells, in particular, could sometimes be killed while surrounding healthy tissue survived. After 1900, however, when doctors learned that treatment in prolonged single sittings might be harmful, they shifted to spaced, briefer sessions. That meant weeks might pass before they saw the reddened skin assumed to show efficacy. Thus they needed some measure of X rays at the time of treatment in terms of delayed effects.[7]

Chemistry offered one approach. Certain substances changed color when exposed to X rays. The changes, roughly matched to later effects, became an index of dose at the time of treatment. Early chemical methods were none too precise—substances might also react to factors other than the rays—but they were useful, simple, and seemed to reflect the biological effects themselves. They remained methods of choice in the clinic for decades. Physics, in contrast, favored an electrical approach. X rays ionized the air or other gases they traversed. Within certain limits, current produced by ions measured degree of ionization, number of ions, and, indirectly, the intensity of the rays. Such methods held little appeal for X-ray workers. Subject to their own sources of error, they were not clearly better than other methods before World War I. Practitioners found the special apparatus awkward to use, the required computation irksome. Ionized gas also seemed far removed from the real concern, biological effects. Only after the war, and partly because of wartime developments, did ionization methods prevail.[8]

The first major change just preceded the war. Physicist William D. Coolidge of General Electric unveiled a new X-ray tube in 1913. His tube expelled electrons from the cathode with heat. Thus needing no gas as a source of ions, it was also a vacuum tube. Teamed with a new power supply marketed a few years before, the Coolidge tube solved the problem of erratic output. In time, it also sharply increased the strength of X rays used for treatment. This effect, however, merely accelerated a long-term trend. Questions about the dangers of increased output surfaced even before the Coolidge tube had attained wide use. Questioning grew more pointed as the number of X-ray workers also climbed steeply, a direct result of wartime demand.[9]

Military planners had early noticed the value of X rays for dealing with bullet wounds and other war-related injuries; machines, in fact, were in the field before 1900. The vastly larger scale of World War I, however, vastly expanded the demand for X-ray services. Hasty training, jerry-built machines, slipshod practices—hallmarks of wartime pressure—clearly augmented the hazards, but they also inspired needed change. Larger numbers of X-ray workers received training that more often included standard safety measures. Radiology emerged from the war with enhanced status and further profited from an influx of academically trained physicists. Some worked for such national standards laboratories as the American National Bureau of Standards, which first became active in radiation protection and measurement during the war. However they entered the growing field, their combined impact sharply altered the context of radiation safety.[10]

Medicine and physics united sooner in Germany than elsewhere.

Between 1914 and 1924 that union diverted the dose question into new channels. Two steps were crucial: first, a well-defined statement of major sources of error in ionization measures; second, proof that biological effects depended only on energy absorbed. What caused the effects remained unclear, but absorbed energy could be assumed proportional to ionization. Ionization could thus be adapted to a biological unit, the unit skin (or erythema) dose. Users need only measure what level of ionization reddened skin in a defined time. Other effects could then be expressed as ratios of this unit dose. Skin response was well known, chambers need not be standard, and clinics might compare results. The new unit quickly achieved widespread use. Physicists objected on the grounds that ions produced physically should be measured in absolute physical, not relative biological, units. Before this view prevailed in the postwar decade, renewed concerns about safety took center stage.[11]

Concern heightened soon after the war ended. Once again, public outcry provoked by news reports was a crucial factor. "The actual stimulus," British medical physicist Sidney Russ recalled, "was the number of deaths among radiologists due to over-exposure."[12] Although known for nearly two decades, deep effects like cancer and blood disease seemed too odd and rare to be of much concern. Safety meant guarding against the kind of acute exposure that caused severe burns, and such safeguards were assumed to protect workers from deep effects as well. The seeming rash of deaths reported early in the postwar decade challenged that belief. Chronic exposure to lower levels of radiation posed dangers that safety measures geared to higher-level acute exposure failed to control.[13]

X-ray workers themselves still bore the major risk, and their own professional societies first addressed the problem. In 1921 the American Roentgen Ray Society formed a standing committee on X-ray protection. That same year, across the Atlantic, "a group of persons really concerned to arrest the sequence of casualties to X-ray and radium workers" founded the British X-Ray and Radium Protection Committee.[14] Other countries followed suit. Each group couched its first recommendations in broad terms, the same pattern of informed common sense that had marked response to hazards for two decades. The essence of their message was caution in using X rays. Little more was yet possible. Although the danger was clear, how to surmount it was not. Until the question of dose was settled, safety standards could not move beyond truisms.[15]

The common sense of accepted safety practices rested on largely untried foundations: the belief that X rays, like other agents that affected living systems, had a threshold of action. Their effects compared, as German doctor and physicist Hermann Wintz explained in 1931, with "the action of poisons. In toxicology the most important thing to know is what quantities are harmful and what . . . can certainly be tolerated . . . without ill effects. Such standards also apply in the case of radiation."[16] Practitioners believed they could find a harmless dose, or at least one so low as to permit cell, tissue, or organ to repair itself. That often tacit belief remained untested by any real attempt to define such a dose quantitatively. The postwar decade saw the first concerted efforts to assign numbers to tolerance, with the new medical physicists playing the central role.[17]

THE CONCEPT OF TOLERANCE

Arthur Mutscheller made the first useful attempt to define a tolerance dose. Initially presented at the 1924 meeting of the American Roentgen Ray Society, his paper appeared in print early the next year. Like so many others during the early 1920s, he felt that "the results of inadequate protection against the harmful effects of over-dosage with roentgen rays . . . are so appalling that a search for protection standards is one of the most important problems in roentgen-ray physics and . . . biology."[18] Mutscheller had come to the United States as chief physicist for a New York maker of X-ray equipment. Born and trained in Germany, he may have known more about recent German research than most Americans, but the problem of protection standards was also in the air. Independently, for instance, Rolf M. Sievert's work at the same time in Sweden closely paralleled Mutscheller's. But it was Mutscheller who triumphed over his own leaden prose, Germanic style, and skimpy data to define the basis of safety standards for the next two decades and more.[19] His paper is thus worth a close look.

Mutscheller posed four linked problems: What factors affected dose? How could the amount of shielding required be computed? What dose could workers tolerate? How could the dose that caused an observed biological change be computed? He dismissed at the outset any notion of absolute safety. Even had theory not rejected it, "the protective shields and apparatus would . . . become entirely too heavy, unwieldy and costly. . . . [W]e must deal with an equilibrium between the amount of protection obtainable and the weight and cost of the protective shields." Achieving proper balance between safety and cost meant every X-ray worker "must be content to receive within a given time, a certain quantity dose of radiation." Biological studies would have to show "just what that dose shall be to which the average operator can be exposed without danger to his health." Required shielding would then have to be computed from the dose so found.[20]

Mutscheller showed how to compute X-ray intensity at a given distance from an unshielded tube. This figure became the basis for finding the erythema dose at that distance. The ratio of tolerance to erythema dose gave the fraction to which shielding must reduce X-ray intensity for the sake of safety. There were two distinct figures: first, the dose a worker "would receive at the given distance from the tube"; second, the amount of shielding required to lower that dose "so that the dose to which the operator is actually exposed is not large enough to injure his health." Still a third figure was needed before the method could be applied: "the dose which an operator can, for a prolonged period of time, tolerate, without ultimately suffering injury." And so Mutscheller came to the tolerance dose.[21]

Mutscheller computed rather than measured his proposed dose. He provided few details: he had surveyed "several typical good installations" and taken "fair averages" of monthly worker exposure. What he meant by these phrases must be inferred from comments elsewhere in the paper. "Good installations" seems to mean those in which workers showed no ill effects from the doses they received. "Fair averages" implied average times to expose a plate, conduct an examination, or complete a treatment session; it also meant the average number of times a day each of these chief kinds of X-ray work was performed. With these figures and known data on machines and shielding, he found the accumulated "dose to which the operators are now exposed during . . . one month." This furnished the basis for tolerance.[22]

Given current practices and standards, Mutscheller judged a worker "entirely safe" who "does not receive every thirty days a dose exceeding 1/100 *of an erythema dose.*" From what we now know, he added, "this seems to be the *tolerance dose* for all" X-ray work. Mutscheller freely admitted that further research might raise or lower the limit; his was merely a starting point. But setting any tolerance dose promoted worker safety "without . . . too heavy, clumsy and unwieldy . . . screens and shields."[23] Mutscheller then showed how his method applied to real cases. The exact thickness of lead required to reduce the computed unshielded dose to the tolerance dose was read from a table based on the fraction of X rays lead absorbed at various wavelengths. If cost, weight, and use permitted, shielding might be increased, but the table gave "the permissible minimum thickness of protective material." Thus were cost and safety balanced.[24]

Mutscheller's approach won strong support from other physicists. A decade elapsed, however, before the concept of tolerance became the formal basis for safety standards. Most doctors disliked the concept and the calculation it entailed. In any event, the question of dose itself still required settling. The effort to do so expanded in 1925 when the first International Congress of Radiology convened in London. Although doctors who attended were mostly content with the unit skin dose as a standard, physicists promoted an ionization-based unit and carried the day. Which unit, however, required more study. The congress formed a committee to report back at its next meeting, scheduled for Stockholm in 1928. The 1928 congress adopted the roentgen, measured by the ionization of air, as the international X-ray unit. Certain details took longer to resolve, but that settled the basic issue.[25] Technically, the roentgen was a unit of exposure, not dose, but for the next three decades most workers ignored this subtle distinction.

Although the 1925 London congress discussed safety standards, it took no formal action. Proposed standards for X-ray shielding lacked a basis in theory. Mutscheller's work, published only shortly before the congress convened, provided the needed rationale. Other studies followed his lead and reached the same results. By 1928, when the Stockholm congress met, most physicists in the health field accepted 1/100 of an erythema dose per month as the tolerance dose. This position was reflected, but only implicitly, in the first internationally adopted safety standards.

The Stockholm congress formed the International Committee on X-Ray and Radium Protection (later the International Committee on Radiation Protection and, despite further changes in name and organization, still known as ICRP). The committee based its first recommendations on the 1922 British-proposed standards, and like them, chiefly addressed good practice.[26] The recommendations included a table of lead thicknesses required to shield workers from X rays produced at given voltages. Although the figures had been adjusted to the tolerance dose, the committee offered only a working guideline: "It should not be possible for a well rested eye of normal acuity to detect

in the dark appreciable fluorescence of a screen placed in the permanent position of the operator."[27]

The ICRP's American member was Lauriston S. Taylor, who had
joined the National Bureau of Standards in 1927 with a B.A. in physics
from Cornell University. The United States had no central protection
group in 1928, a lack Taylor acted to remedy when he returned from
Stockholm. The result was the Advisory Committee on X-Ray and Radium Protection, which first met in September 1929. Like ICRP, the
American committee has changed its name and structure; now the National Council on Radiation Protection and Measurements, it is commonly referred to as NCRP through all its manifestations. The two
major American professional groups—the American Roentgen Ray Society and the Radiological Society of North America—each sent two
members, one a physicist, the other a radiologist. They were joined by
a spokesman for the American Medical Association, and two members
chosen by the X-ray equipment manufacturers. Although supported by
the National Bureau of Standards, the committee had no formal standing.[28] It "carefully avoided the question of to whom the Committee
might be advisory," Taylor later explained. "Actually, it felt itself to be
responsive to the profession as a whole and clearly avoided any implied direct responsibility to either the government or to any individual
radiological society."[29]

The committee's first goal was to write a report on the subject of
safeguards against X-ray hazards. Entitled *X-Ray Protection*, it appeared
in May 1931, just before the international meetings in Paris. Despite
the seeming imprimatur of the National Bureau of Standards, which
issued the report as NBS Handbook 15, its recommendations carried
no legal force. Yet it looked official. Many must have thought that NBS
Handbook 15 and other reports issued as bureau handbooks over the
next three decades did express government policy. The 1931 report, in
any event, demanded little of those who might wish to comply. It retained the same pragmatic stress on shielding tables and ignored the
question of tolerance dose.[30] In essence it remained merely, in Taylor's
words, "a very greatly expanded version of the 1928 recommendations
of the International Committee."[31]

ICRP discussed the problem of tolerance dose at the 1931 Paris meetings. Since the standard unit was adopted in 1928, tolerance values
had been converted from fractions of an erythema dose to roentgens.
Some workers expressed the value in seconds or days rather than

months, but the dose itself remained unchanged. Research equated the erythema dose to roughly 600 roentgens; the tolerance dose then became 6 roentgens monthly, 0.2 roentgen daily, or 0.00001 (10^{-5}) roentgen per second. The expanded shielding tables ICRP recommended in 1931 were computed from a tolerance dose of 10^{-5} roentgen per second. Once again, however, tolerance was not mentioned. Many doctors still disliked the approach, which seemed too abstract and mathematical; they preferred erythema dose as more closely related to the biological realities. Then, too, the figures were not really equal from their viewpoint. Physically identical doses could produce quite varied effects in living systems; individual response depended on dose rate, total time of irradiation, time between exposures, and many other factors.[32]

The American Advisory Committee adopted the first explicit value for tolerance. Its second report, issued in March 1934, suggested 0.1 roentgen per day: "The safe general radiation to the whole body is taken as 1/10 r per day for hard X-rays and may be used as a guide in radium protection." No reason was offered, and a warning followed: "Too great reliance is not to be put on the above figures."[33] Members of the committee had discussed the subject at length. Quickly united in strong support for setting a tolerance dose, they hesitated over the precise figure. Mutscheller's value converted to roentgens furnished the starting point. Practicality dictated a daily dose. Based on the normal workweek, this procedure yielded 0.24 roentgen per day. As that number looked too precise, implying more knowledge than the facts warranted, members agreed to round the figure downward. Because they decided only after the report was in press, the figure appeared as a note under "Personnel." In the committee's third report, the revised *X-Ray Protection* issued as National Bureau of Standards Handbook 20 in July 1936, the tolerance dose became part of the text.[34]

Meanwhile, at Zurich in 1934, the ICRP also acted. It set the daily dose at 0.2 roentgen, with a direct but cautious statement:

> The evidence available at present appears to suggest that under satisfactory working conditions a person in normal health can tolerate exposure to x-rays . . . of about 0.2 . . . roentgens . . . per day. . . . The protective values given in these recommendations are generally in harmony with this figure under average conditions.[35]

The ICRP and the American group had taken the same path, Taylor explains, "a crude evaluation and summation of all of the . . . results" since 1925.[36] Each group rounded its figure downward, the ICRP simply not as far. Even the higher figure seemed conservative, based as it was on the absence of any observed ill effects at somewhat higher doses.[37]

By the mid-1930s the concept of tolerance had largely prevailed. Standards were expressed as the thickness of lead shielding required to prevent workers from being exposed to doses above tolerance. The absence of observed damage implied safety. Such standards served for other toxic agents and, in practice, also worked well enough for radiation. The research basis for the concept, however, remained quite limited; no proof of tolerance existed. Yet even while winning its widest support the concept was being challenged. Questions came from several quarters. Perhaps the most telling concerned the radium problem. The same ICRP statement that set the tolerance dose for X rays in 1934 concluded: "No similar tolerance dose is at present available in the case of radium gamma rays."[38]

THE HAZARDS OF RADIOACTIVITY

All the protection groups of the 1920s included radium as well as X ray in their titles. Prompted by Röntgen's work, Henri Becquerel in Paris found in 1896 that uranium emitted invisible rays. Thorium, polonium, and radium joined the list of active substances in 1898; actinium and radon soon followed. Marie and Pierre Curie, Becquerel's colleagues, discovered both polonium and radium. They also coined the term "radioactive" and gave their name to the unit of radioactivity adopted in 1912. Radioactivity comprised three kinds of rays; Ernest Rutherford labeled them alpha, beta, and gamma. By 1903 he and Frederick Soddy explained the rays in terms of energy emitted when atoms of one element transmuted to a second. All three came from the nucleus, but only gamma were true rays. So-called alpha and beta rays proved to be streams of particles. Each active substance displayed its own characteristic mode of decay and energy distribution. Emitting gamma rays and either an alpha or a beta particle at each step, radioactive elements passed through a fixed series of changes at a constant rate ending in a stable isotope of lead. During the next decade physicists and chemists explored links between the natural radionuclides. By 1913 they had largely defined the three main decay series: in modern terms, from uranium 238 through radium 226 to lead 206; from thorium 232 through radium 228 to lead 208; and from uranium 235 through radium 223 to lead 207.[39]

Well before these relationships were fully explained, however, radium joined X rays as a useful tool and sometime hazard. Once known to have effects like those produced by X rays, radium also quickly found medical uses. But so rare and costly a substance—the first gram of American radium in 1913 was refined from 500 tons of ore and cost $120,000—could serve only limited uses and endanger few workers. The danger, in any event, seemed modest, easily avoided by careful handling and proper shielding. Such was the substance of both ICRP and American Advisory Committee recommendations. These measures sufficed to cope with the skin burns that long seemed the major threat. Unfortunately, radium gamma rays, like X rays, also produced deeper effects, slower to appear and more severe.[40]

The problem emerged in the early 1920s, just when concerns about deep effects from X rays also revived. Reports appeared of blood disease in workers long exposed to gamma rays. Chronic exposure damaged bone marrow, the body's main blood-forming organ, with sometimes fatal results. Leukemia might also be a nasty sequel to prolonged exposure. Even a severe dose might take years to produce its effects. In this respect, gamma rays posed hazards distinct more in degree than in kind from X rays. Although produced elsewise, gamma rays are the same as very hard (high-energy) X rays; the International Committee on Radiological Units in 1937 redefined the roentgen to include gamma rays.[41] As Robley D. Evans remarked in his 1933 review, "Over-doses of gamma rays will produce pernicious anemia and myeloid leukemia as late as 4 years after a massive exposure, but these effects are to be regarded as radiation burns rather than as true radium poisoning."[42] Radium as poison had no parallel in X-ray usage.

The first published report of a human being poisoned by radium came in 1924. A young woman from northern New Jersey was referred to Theodore Blum, a New York dentist and oral surgeon, when her jaw failed to heal after dental work. It was inflamed and showed signs of necrosis: the bone was dying. Blum knew she had worked at painting luminous figures on dials. He also knew the plant contained radium. Attributing her symptoms to taking radium into her mouth, he labeled the syndrome "radium jaw."[43] Harrison S. Martland, a county health official in New Jersey, saw Blum's note. Investigation quickly confirmed Blum's guess and revealed the nature of the problem. Radium is chemically akin to calcium. Although painters used insoluble compounds, their bodies did absorb some radium. Fixed in bone, it bombarded both bone and marrow with alpha particles. Rotting bone promoted infections; dying marrow failed to maintain the body's blood supply. Either sepsis or anemia could be fatal. Even if a victim survived these relatively quick effects, cancer might follow. How dial painters came to be poisoned by radium was a lesson in the dangers that may attend early use of a poorly known substance.[44]

Just before World War I the supply of radium sharply increased, and commerce expanded when prices dropped. Expensive watches and other instruments began to feature dials with self-lit figures. The effect was achieved by painting the figures with a mixture of zinc sulfide and tiny amounts of radium, at most one part radium to 40,000 parts sulfide and binder; that was just enough radium to cause the zinc sulfide to fluoresce. War spurred demand for dials that glowed in the dark; production soared from under ten thousand in 1913 to over two million in 1919. As painting by hand was the rule, the effect was to multiply so-called studios rather than increase plant size. The largest belonged to the United States Radium Corporation, founded in 1916 as Radium Luminous Materials Company. Located in Orange, New Jersey, it never employed as many as three hundred workers at one time. The industry as a whole, which was centered in northern New Jersey, may have had a work force upward of two thousand, almost all young women.[45]

To sustain itself after the war, the industry sought new markets. It successfully promoted luminous doorknobs, light switches, and other novelties, as well as low-priced watch and clock dials. United States Radium pioneered two cost-cutting moves in shifting to a mass market. Mesothorium, cheaply available as a by-product of the local thorium gas-mantle industry, replaced or supplemented radium. Actually, this procedure simply substituted one isotope of radium for another. Radium, so-called, was radium 226, derived from uranium 238. Mesothorium was radium 228, produced when thorium 232 decays by emitting an alpha particle. The rate of decay is constant, most often expressed as half-life, the time required for half some number of atoms to transform themselves. Mesothorium's half-life is six years; that of radium 226, more than sixteen hundred. Inexpensive timepieces hardly required the longer-lived and far more costly isotope.[46]

The company also replaced oil-based with cheaper water-based paints. Oil-based paints were applied with glass rods or metal pen nibs, but water-based paints required fine brushes. Workers pointed the brushes with their lips, a practice known as tipping. Inevitably, they got paint in their mouths and swallowed some. How much was ingested depended on each worker's habits and skills, which varied widely. In his 1933 review Evans judged that an average worker might have "licked 1 mg. of paint from her brush 4 times per dial, 300 dials per day, 5 days per week,"[47] or about 6 grams of paint a week. At one part per 40,000, the radium total might reach 150 micrograms (0.00015 gram). Some estimates were a bit higher, others even lower, as little as 7.5 micrograms per week. Initially, no one dreamed that such minute amounts posed any hazard at all.[48]

But they did. By late 1923 United States Radium warned workers against tipping their brushes. Early the next year it commissioned a survey of its plant by a team of occupational health experts from Harvard University. They found radioactive contamination severe enough to constitute a serious health hazard. Facing litigation, however, the company refused to publish the findings for well over a year.[49] Meanwhile, local dentists were talking among themselves about the "jaw rot" that seemed to afflict dial painters. Blum's note linking jaw necrosis to radium was merely the first public statement of suspicions long harbored. By the end of 1924 at least nine young women who had worked for United States Radium were dead. There may well have been more. Symptoms could take years to appear and the syndrome was unknown.[50] Doctors, as Martland remarked, "might treat a victim for sepsis, anemia, Vincent's angina, rheumatism or 'God knows what.' "[51]

The situation changed in 1925, when Martland, chief medical examiner of Essex County, New Jersey, began his investigation. He examined several painters and autopsied two who died. Tests showed the bones highly radioactive. Martland could also detect radon in the breath of living subjects. Often termed radium emanation, radon 222 has a half-life of four days; itself an alpha emitter, it arises from alpha decay of radium 226. As a relatively inert gas, it diffuses through the bloodstream, enters the lungs, and appears in exhaled air. So does thorium emanation, or thoron (radon 220), a decay product of the mesothorium used in the paint mixture. Martland regarded either gas as a clear danger signal, even lacking other symptoms. Of one young woman who seemed to be in perfect health, he commented that "she could at the present time pass a strict medical examination and obtain almost any amount of insurance." Radon on her breath, however, meant "radium and mesothorium in her bones, spleen and liver, . . . constantly bombarding her blood-forming centers. Who can tell when she may develop an acute fatal anemia, or a more chronic anemia, with or without local lesions and bone necrosis?"[52] Four years later Martland reported her "a victim of chronic bone lesions of a crippling nature."[53]

By then strict controls on tipping had long since been imposed. The familiar pattern of headlines, public outcry, and legal action once again forced drastic changes in practice. That was scant help for women already exposed; no one knew how to remove radium from their bodies. Necrotic bone and blood disease proved only early sequels. In 1927 Martland autopsied another victim and found bone cancer; she carried 50 micrograms of radium in her skeleton.[54] As Evans noted in his 1933 review, "when fixed in bone, as little as 2 micrograms has been fatal."[55] Although the body retained only 2 percent of radium ingested, the total accumulated. Just how much any worker swallowed was never certain; not only did habits vary, but each company guarded its paint formula as a trade secret. Even at the lowest estimated intake, however—7.5 micrograms of active substance per week—three months of steady work might produce a lethal burden.[56] In time cancer became a leading cause of mortality among women who had worked as dial painters before 1925.[57]

Despite the headlines, dial painters were not the only victims. A second group comprised the chemists and other workers who extracted radium from its ores or prepared its compounds in the laboratory. Inadequate ventilation often exposed them to active dust or radon. Careless technique—using a mouth pipette to transfer active solution was a common practice—could result in swallowing radium; so could accidental spillage. Just such sources were implicated in two cases Martland reported from United States Radium. Although neither man had ever painted a dial, both suffered the same syndrome as the dial painters and both died. One revealed at autopsy a skeletal burden of 15 micrograms. Perhaps the largest class of victims, however, comprised people dosed with radium for medical reasons. Their plight provoked a new round of concern about radium hazards in the early 1930s.[58]

PROTECTION STANDARDS FOR RADIUM AND RADON

The American Medical Association approved radium for internal use from 1914 until 1932. Elsewhere it remained an accepted treatment even longer, in Germany until 1952. Thousands of patients swallowed or were injected with dissolved radium, as Evans remarks, "for any medical condition which had no known cure, including arthritis, neuritis, hypertension, poliomyelitis, menopausal complaints, Hodgkin's disease, debutante's fatigue and even dementia praecox."[59] As radium commerce expanded, its glamour was also exploited in home cures. Tonics and nostrums purchased over the counter became faddish. Many were largely harmless, but some were not. One tonic (and reputed aphrodisiac) widely used by those who could afford it was sold in cartons of thirty vials. Vials contained one microcurie each of radium and mesothorium in half an ounce of water, one vial to be taken after every meal and one before bedtime. Prominent Pittsburgh industrialist and man-about-town Eben M. Byers became a faithful user. Headlined nationwide, his gruesome death in 1932 provoked a public outcry.[60]

Until then, what amount of ingested radium might be harmful was more discussed than studied. The Byers case proved catalytic. In southern California, local health authorities decided to crack down on radium quacks. The California Institute of Technology provided help in the person of Robley Evans, then a graduate student using very sensitive techniques to measure radium and radon. The puzzle of turning such techniques to finding the radium content of people engrossed Evans. Soon after joining the Massachusetts Institute of Technology faculty in 1934, he embarked on a long-term project to define safe body burdens. The Food and Drug Administration partly sponsored the work, concerned to know how much radium to allow in a growing range of commercial products like face creams and vaginal jellies.[61]

How to measure from outside the exact amount of radium fixed inside a living person's body was the key problem. There was no direct method, since the body itself blocked the rays. Radium decayed to radon, however, and exhaled radon clearly signaled someone burdened with radium. That posed a new question: What was the ratio between exhaled radon and fixed radium? The MIT studies began with rats, but valid human doses could not be derived from doses administered to animals. This problem led Evans and his co-workers back to the dial painters. By 1941 they had accurately measured the body burdens of twenty-seven such workers. Seven bore less than 0.5 microcurie and showed no ill effects. The other twenty all had burdens larger than 1.2 microcuries; all suffered some degree of damage. These findings supported the first published figures on permissible levels of radon in the breath and matching radium body burdens.[62]

Publication followed a meeting called by the National Bureau of Standards in February 1941. Its purpose was to set standards for the safe handling of radioactive luminous compounds. Although widely discussed for years, the matter became urgent with the outbreak of war in Europe and surging military demand for luminous dials. Presented with a draft text, Evans and his colleagues quickly agreed to proposed housekeeping rules, patterned on normal good practice. What really mattered was which numbers should fill the blanks left for tolerance levels. The MIT findings proved decisive. Evans reviewed the twenty-seven cases. Then, as he recalls the scene,

> noting that we were obliged to make an "informed judgment" decision, I suggested [setting] the "tolerance level" for residual radium burden . . . at such a level that we would feel perfectly comfortable if our own wife or daughter were the subject. I then asked each of the other 8 committeemen individually in turn if he would be contented with 0.1 μCi. Unanimously, they all were.[63]

This standard was published in National Bureau of Standards Handbook 27 of May 1941.[64]

The same handbook also set the first standard for inhaled radioactive gases. It, too, derived from MIT studies. In 1937 the insurer of a plant making thorium gas mantles requested MIT help in settling questions about radiation hazards in the plant. The problem centered on thoron (radon 220). Inventing their own air-sampling techniques, Evans and his colleagues found that some parts of the plant contained levels of gas they judged dangerously high on the basis of recent European data. Miners in Saxony and Bohemia had long been known to suffer abnormally high rates of lung cancer. During the 1920s researchers traced the cause to radon released from ores in the mines. Levels as low as 0.000000001 curie (10^{-9} Ci) per liter of air presented a hazard. Applying a safety factor of 100 to this figure, Evans in 1940 proposed "10 μμ Ci (10^{-11} Ci) per liter as the safe working concentration of radon or thoron in plant, laboratory, or office air." This figure was adopted in the 1941 handbook.[65]

The 1941 standards completed a structure of safety founded on the concept of tolerance. By 1941 the hardest questions concerned that concept itself. Biological effects of radiation had received little study before World War II. Genetics was a partial exception. If tolerance meant anything, it must be that radiation produced no ill effects—at least none beyond the power of a cell or other living system to repair itself—until reaching a certain level. Biologically, such thresholds are common; few supposed that X rays would differ from any other toxic agent in this regard. But from 1927 on, when Herbert J. Muller published his first results on mutations induced by X rays in fruit flies, growing evidence challenged this widespread assumption. The direct proportion between amount of radiation and number of mutations seemed without lower limit. Absence of a threshold for this effect might mean the same for other effects. What looked like tolerance could simply reflect effects too subtle for current techniques to detect. Whether unique or not, however, the so-called threat to the germ plasm seemed dire enough in its own right to demand action from the American Advisory Committee.[66]

Genetic concerns induced the members at a meeting in December 1940 to propose a lower tolerance dose for the revised handbook on X-ray protection. The new figure was 0.02 roentgen.[67] They soon had second thoughts. A member who missed the December meeting, Gioacchino Failla of New York's Memorial Hospital, was an expert on radium protection. He objected strongly to the lower limit. The risk of "damage from .1 roentgen per day is so slight that one can just as well stop at this point." A lower dose, he argued, would simply hinder using radium without adding much to safety.[68] Failla's case persuaded the other members. Meeting in September 1941, they decided to postpone action until further research pinpointed the hazard.[69]

Meanwhile, though, word of the proposed change got out. Ordinarily, committee recommendations became public only when published by the National Bureau of Standards. In this instance, however, the normal process was short-circuited. Taylor himself deserved part of the blame. In a 1941 article he referred to a recent review finding "that a person may receive without injury up to one-fiftieth roentgen daily." Taylor did not, in fact, endorse this figure, but neither did he cite any other.[70] With the question much in the air, the hint sufficed. Taylor's very cautious remark was more than once cited as authority for a new tolerance dose.[71] Part of the problem, as the committee noted at its September meeting, resided in the very term "tolerance dose." It implied a dose that

> can be tolerated without any damage whatever, which, of course, is not the case if we consider genetic damage. It was recommended therefore that in the future we use the term "permissible dose." This does not in any way imply that no injury will follow. It merely says that the Committee recommends its use even though it is not necessarily safe, but is adopted only as a practical and expedient value.[72]

The matter went no further, however, as this meeting proved to be the committee's last. When the committee was revived after World War II, both its name and context had been transformed. Ironically, the leaders of prewar standard setting played no major roles in the wartime project that posed the greatest challenge to radiation safety, the development of atomic bombs.

Capping a decade that transformed nuclear physics had come the discovery, announced in January 1939, of nuclear fission. Uranium atoms struck by neutrons split to form lighter elements. In the process, mass became energy. Even before World War II broke out later that year, alarmed refugee scientists from fascist Europe warned of the danger. If a fissioned atom produced enough neutrons, then a large mass of uranium might support a very rapid chain reaction. The result could be a bomb of immense power. Albert Einstein urged the president of the United States to sponsor research aimed at learning whether or not such a bomb could be made. President Roosevelt endorsed a modest program divided among several university laboratories. July 1941 marked a turning point. A British report concluded that a bomb might be produced quickly enough to affect the current war. Meanwhile came word from Berkeley that the newly discovered element 94 (later named plutonium) promised to support a chain reaction. At a White House meeting in October, although the United States was not yet at war, President Roosevelt ordered bomb development to proceed. The heart of the job was proving that a bomb would work. Showing that a chain reaction, still only theory, could be achieved in practice became the crucial first step. This task was entrusted to the so-called Metallurgical Project at the University of Chicago.[73]

The Metallurgical Laboratory (the Met Lab as almost everyone called it then and since) conducted the project. Created in January 1942, it was led by Arthur H. Compton, winner of a Nobel prize in physics and long-time member of the Chicago faculty. Newly arrived from Columbia University where he settled after fleeing fascist Italy, Enrico Fermi, also a Nobelist, took charge of the Physics Division. The goal was building a nuclear reactor, then termed a pile—literally a pile of graphite blocks, stacked to form a lattice holding slugs of uranium. Uranium provided the neutrons which graphite slowed enough to allow a controlled chain reaction to proceed. Such was the theory, at any rate, that the pile would confirm. If it worked, then later and larger versions would supply the makings of a bomb.[74]

The choice of bomb materials was restricted. Only two seemed likely to support a rate of fission suited to an explosive chain reaction. One was uranium 235. Unfortunately, that isotope accounted for less than 1 percent of the total metal in natural ores composed chiefly of uranium 238. Using uranium required learning how to enrich the mix, that is, to increase the ratio of uranium 235 to uranium 238. Identical chemically as the two were, the method must be physical. Practically, that meant finding how to exploit the very slight differences between their weights. Several techniques held promise, but the large plants built to apply these processes posed only minor radiological hazards. Uranium was simply not active enough to threaten workers from that standpoint.

Plutonium, the other prospect for bomb use, was something else. Producing it loomed as the Met Lab's hardest job. Neutrons within a chain-reacting pile converted some uranium 238 fuel to plutonium 239 as well as making a mixed stew of fission products. From this mixture the plutonium was then chemically extracted. Every stage of running a pile and handling its products posed radiological hazards. Their exact nature and extent were questions; potentially, however, they clearly far surpassed any known in the past. This unique challenge explains the early and lasting stress on radiation safety which marked the Chicago project. In time it made the Met Lab's Health Division a major source of safety practices for the bomb project as a whole.[75]

Much remained to be learned, but the problems' broad outlines emerged clearly enough. No less crucial, means of dealing with them lay ready to hand. The concept of tolerance and working methods for coping with radiation hazards devised in the prewar decades furnished the safety framework for the wartime project. Merril Eisenbud has commented on "the remarkable historical coincidence" that the needed standards existed just when they had to.[76] Published tolerance standards for X rays appeared little more than seven years before the Chicago project began. Radium and radon standards were scarcely seven months old. Precise figures were in some sense arbitrary. No one could prove the safety of 0.1 roentgen of gamma or X rays, 0.1 microcurie of radium, or 10 picocuries of radon per liter of air. A careful person might say only that someone exposed to that dose, or who swallowed that amount, or who breathed air at that level, appeared to be unharmed after a certain length of time. Standards reflected best, not final, judgments and were offered with due care, often qualified by phrases like "according to present knowledge."[77] Intended to provide working guidelines in a field still marked by many questions, they served well.[78] The informed judgment of prewar experts facing hard choices and limited data largely proved itself in the crisis.

FORMING THE HEALTH DIVISION

The Chicago Health Division began modestly. While Compton was creating the Met Lab in the first months of 1942, two physicists visited Billings, the University of Chicago's medical teaching and research hospital. They wanted someone to tend health problems for "a project . . . on campus." The choice was Leon O. Jacobson, a young hematologist two years out of school. Awed in the presence of a dean, the head of medicine, and the physicists, Jacobson heard about "the importance of whatever it was they were doing, but [not] what it was beyond some vague reference to cyclotrons."[79] His task seemed routine—arranging for physical examinations of project workers—except for an unusual stress on skin tests, blood counts, and urinalyses. Physicists' knowing that radiation caused blood changes was the reason for seeking a blood expert as liaison between project and hospital. Hoping to keep close track of blood data from all workers as a means of spotting exposed workers early, they were disappointed. Normal blood counts varied so widely, or were affected by so many factors, that the technique proved not very useful during the war.[80]

Establishing a clinical screening program, however, was only a first step. Much more had to be learned about the biological effects of radiation. The need seemed urgent as Fermi's progress during early 1942 strongly suggested the pile would soon be running. Compton recalled, perhaps with a touch of hyperbole, that

> Our physicists became worried. They knew what happened to the early experimenters with radioactive materials. Not many of them had lived very long. They were themselves to work with materials millions of times more active than those of these early experimenters. What was their own life expectancy?[81]

The physicists decided they needed a biophysicist to lead a research program.

To those working at the Met Lab who knew him, Kenneth S. Cole seemed just the man for the job. Trained in physics and a professor of physiology at Columbia University, he had pursued biophysical research for nearly two decades. Chicago first contacted Cole in spring 1942 with an informal invitation to join a project about which he could be told little. Hearing nothing more, however, he accepted another war research job in Washington. Then came an urgent call from Compton himself. A clearance as consultant had been rushed through, the project "needed help, and would I please come to Chicago as soon as possible." Cole left the next day. Although excited by the research prospect, he declined to head the nascent Health Division. That would entail medical responsibilities for which he regarded himself unqualified.[82]

Continuing his search, Compton found Robert S. Stone. Canadian-born, Stone trained in radiology at the University of Toronto, but he had taught at the University of California Medical School since 1928. Recently named department chairman, Stone enjoyed close ties with pioneering cyclotron work at the university's Radiation Laboratory. That placed him in the forefront of medical research on the uses and effects of supervoltage X rays, radioisotopes, and neutrons, all bearing directly on hazards most likely to confront the Chicago project.[83] "Stone's exceptional qualifications for this work on which the very lives of our workers depended was evident," Compton recalled.[84] Stone took charge of the newly formed Health Division on 6 August 1942. Cole, already on the scene, headed the division's biological research section.[85]

By curious chance, Stone also found another radiologist already in Chicago. As Compton sought the advice that led to hiring Stone, Simeon T. Cantril was stopping in Chicago to visit old friends. He had spent five years in the city during the 1930s at Michael Reese Hospital and the Chicago Tumor Institute. Since then he had directed the Tumor Institute at Seattle's Swedish Hospital. Now the army had called him for routine medical services. The Chicago project could clearly make better use of his special talents; he took charge of the division's medical section. Leon Jacobson, who had run the screening program since February, now became a full-time member of the Met Lab as Cantril's assistant.[86]

Also on the job when Stone arrived was Ernest O. Wollan, who had trained under Compton. Wollan's prewar research centered on measuring cosmic rays. Compton summoned him to join the Met Lab's Engineering Division, where he formed a group to search for radiation leaks and to compute the amount of pile shielding required. When Stone arrived, Wollan became chief of his health physics section. How and why this odd term came to designate radiation protection has been much discussed. Probably, as one early Health Division recruit suggests, the coinage at first merely denoted the physics section of the Health Division.[87] The term "health physics," however, soon acquired and retained broader meaning as title of the profession concerned with radiation protection. Specialists in this field predated the Health Division but lacked a unique title. Despite such public outcries as those occasioned by the unhappy fate of radium dial painters, radiation safety had never risen to the level of major social concern; no one felt compelled to single out its practice for special notice or label. That the need now appeared may underline the expected impact of the Metallurgical Project's work toward a fission bomb. The name also served security: "radiation protection" might arouse unwelcome interest; "health physics" conveyed nothing.[88]

As its central task, the Health Division safeguarded project workers against the hazards of radiation. The choice of Stone as director and of Cole, Cantril, and Wollan as chiefs of biological research, medical, and health physics sections completed the basic framework in the summer of 1942. Next came the task of staffing the division. The director and the section chiefs sought new people as they themselves had been recruited. The same pattern, in fact, broadly characterized all American war research. Crucial in shaping American science policy, it was controversial then and has remained so. Likely choices for key posts were largely limited to the small group of men personally known to one another, and to their most promising students.[89] Joining Jacobson under Cantril was James J. Nickson, also just two years beyond medical school; as a student he had worked in Cantril's Seattle laboratory. Herbert M. Parker, a noted British radiological physicist and Cantril's colleague in Seattle since 1938, joined Wollan's section. Cole recruited Raymond E. Zirkle, an Indiana University botanist well known for his radiobiological studies, whom he also knew personally. Zirkle led Cole's radiobiology group. To head his biophysics group, Cole added a freshly minted University of Chicago chemist, Richard Abrams. His biochemistry group leader was Waldo E. Cohn. As a graduate student at Berkeley, Cohn had known Stone and others working with radioisotopes.[90]

The process reached the lower levels of the division as well. Charles W. Hagen, Jr., and Eric L. Simmons, for instance, had been Zirkle's students at Indiana. They joined Zirkle's group as did his colleague,

John R. Raper.[91] This method of recruitment was then, and by many still is, deemed the natural and efficient way to secure men of proper skill and sound judgment for important tasks. In the context of a highly secret wartime project, it may have been the only way. No one could be told much more than how interesting the work might be. Willingness to come on that basis testified as much to the personal network as to being invited in the first place: "I would never have gone there [Chicago] except that I had tremendous respect for him [Cole]," recalled C. Ladd Prosser, then a junior professor at the University of Illinois.[92]

Stone and his colleagues had their work cut out for them. Half a year had already passed before Compton decided to add a health division to the Metallurgical Project. Those on the scene had taken the first steps, but a host of urgent tasks faced the new division. It was not merely a question of setting up proper safeguards to ensure worker safety. Too much remained unknown. From the outset, research on the biological effects of radiation ranked high on the Health Division agenda. In part, that research addressed immediate problems of worker safety, but, as Compton recalled, "we were looking also further into the future. What would be the radiation effects of the bombs that we would use? What would be the dangers to which people would be exposed as a result of the everyday use of radioactive materials in an atomic age?" Current problems must be faced, but long-range studies also needed to "be taken in hand now. The results of the studies were sure to become important."[93]

Few if any of those involved in the project dissented from such views. Clearly much remained to be learned about the effects of radiation. What did become an issue was quite a different matter: Did the kind of research required to answer such questions really belong in a wartime crash project under severe constraints of time, money, resources, and manpower? The issue did not arise at once. Strains appeared only when the army took over a project shifting toward development and production. Laboratory research remained the chief focus of the entire bomb project throughout 1942. Radiological safety problems largely conformed to prewar experience, despite much larger numbers at risk. On 26 November 1942 the Health Division issued a reassuring report: Cantril noted the broad familiarity of the hazards; Wollan dismissed any danger from the soon-to-be-running pile; Cole asserted that the problems of fission products could be controlled.[94] A week later, Fermi and his colleagues attained the long-awaited chain reaction. On 2 December 1942, Stone later observed, "the problems of the Health Division" ceased to be "mainly theoretical."[95]

2

ROLE OF THE CHICAGO HEALTH DIVISION

FIRST TASKS

The Health Division issued its first report on 15 August 1942, little more than a week after Robert Stone arrived. It merely noted the need for radiation safety.[1] Serious discourse on what needed to be done and how it should be done appeared only a month later. Simeon Cantril and Kenneth Cole opened the second report with a statement stressing the division's dual role. Part of the job was project health and safety. These concerns were clinically focused; the division bore "direct responsibility for the health of the personnel and the public." The other part was research on "the effects of physical and chemical hazards . . . to provide both a basis for clinical measures and fundamental understanding of the phenomena." These statements in essence defined the goals of Cantril's medical section and Cole's biological research section.[2]

The health physics section supported these efforts. Within the division it measured, analyzed, and computed "the factors involved in the clinical and experimental aspects." Outside the division it had two main jobs: first, to judge and measure future as well as present hazards throughout "the project—research and development, plants and production, and military"; second, to design and oversee safety measures.[3] In practice these statements did not draw sharp limits for the sections; rather they defined centers of concern. Each section did research, just as each became involved in health and safety, guided by the central aim "to eliminate physical and chemical hazards before they reach clinically detectable proportions."[4]

"The first tasks," however, Cantril and Cole noted, centered on "immediate surveys of the hazards to personnel on the Project as a whole."[5] Two especially seemed to require prompt action. Uranium handling was one. Suspected of being highly toxic in animals, the metal was feared so in humans as well. No proven case of a worker poisoned by uranium had yet appeared at Chicago or in plants that supplied fuel for the pile. Yet the Health Division had to assume uranium to be as toxic as other heavy metals, and so it altered handling routines to reduce dust workers might breathe or swallow. At the same time it sponsored research to find how toxic ingested uranium really was. Albert E. Tannenbaum, a cancer expert at Michael Reese Hospital in Chicago, received a contract and began work in September 1942. His findings proved most reassuring. Uranium toxicology later became the subject of independent research at the University of Rochester.[6]

Uranium was not regarded as a radiological hazard. The half-life of uranium 238 was measured in billions of years, of uranium 235, in hundreds of millions. Radioactivity of neither was thus high or intense enough to pose much threat, even if ingested. This points up an aspect of Manhattan Project safety problems easy to miss. As it became a large industrial project using a host of new techniques and substances, most hazards had little or nothing to do with radiation. Workers may well have faced more serious risks over the course of the project in handling toxic chemicals than they ever did from gamma rays, neutrons, or active substances. Radiation was nonetheless the "special hazard." It was the other one of the two earmarked for prompt action in the September report.[7]

Even before heading the health physics section, Ernest Wollan had begun to survey the laboratory for radiation hazards. Leon Jacobson had likewise taken a health role in the project since early 1942. Both believed "some radical changes must be made at once in working conditions to avoid unnecessary exposure to radiation."[8] Cantril began a systematic survey of the health status of project workers. Aided by the staff of the University of Chicago clinics, he began examining several workers a day. They hoped to correlate their findings with each person's past and current work on the project and thus perhaps devise standard blood and urine tests to detect early signs of radiation damage. They were disappointed. Although routine screening of project workers persisted throughout the war, it proved of little use, for normal results varied too widely and were affected by too many factors. Routine testing simply provided no basis for early warning.[9]

Meanwhile, all workers in areas where gamma radiation levels might be troublesome received pocket ionization chambers. Readings were taken and recorded daily.[10] Although a decade old, pocket chambers remained neither handy to use nor easy to buy. A prewar market for radiation detection or monitoring devices of any kind scarcely existed. Ordinarily, users made their own or did without, but pocket chambers were modest exceptions. Victoreen Instrument Co. of Cleveland, Ohio, offered the first commercial model in 1940. It was a simplified version of the well-known Victoreen R-Meter, widely used during the 1930s in X-ray therapy. Victoreen called the new system the Minometer. Commonly, however, workers used the name only for the separate instrument required to charge and read the pocket chamber.[11]

The chamber itself was an air-filled tube the size of a fountain pen. It worked as a capacitor. A thin wire running through the center of the tube and insulated from its walls stored an applied electrical charge. Ions produced in air by radiation neutralized the charge; the difference between initial and final charge indicated how much radiation the chamber had been exposed to. Requiring both charging and reading on a separate device, the chamber was awkward to use. It was also none too reliable because other factors, like shock or moisture, might affect results. Problems in volume manufacturing to Health Division specifications were also slow to be resolved. To sidestep these drawbacks, Herbert Parker suggested giving each worker two chambers. Using them in pairs became standard practice, sharply reducing the ratio of bad readings. Since all errors increased apparent exposure, the lower reading was taken as correct. By 1945 the pocket dosimeter was far more convenient than the chamber. Basically similar to the simple chamber, it still required charging, but its wearer could read a built-in dose scale directly through a system of magnifying lenses. Instruments of this kind became one mainstay of personnel dosimetry throughout the Manhattan Project.[12]

Photographic film was the other mainstay of personnel dosimetry. Pocket dosimeters provided direct monitoring of individual exposures. Wearers could read them at any time and they were recharged daily. Film served a different purpose. It recorded cumulative exposure over days or weeks, although film, too, might be processed on a daily schedule. In September 1942 the use of film was still only in the planning stage for the Metallurgical Project. Dental X-ray film packets had been on the market for two decades; for nearly as long, they had been used to warn workers against pending overexposure. By 1942 the technique was well known if not widespread. Standardization presented problems, as did questions about how photographic emulsions responded to different kinds of rays.[13] Film's value thus remained chiefly qualitative, as one dosimetry pioneer suggested in 1939: "If after a week of exposure and standard development, none of the films is so dark that, when it is put on a printed page in good light, the letters cannot be distinguished through it, then protection is adequate."[14]

Quantification of film-recorded doses, however, made progress. The greatly increased demands and resources of the wartime project promoted still more. Film had drawbacks. Heat or moisture could distort readings. It was more sensitive to some kinds of rays than to others. It could not provide immediate readings, as it had to be processed. But it also promised a simple, cheap, and reasonably reliable way to check the very large numbers of workers who might be exposed.[15] Wollan planned to issue film packets "to all those for which there is any possibility of a radiation hazard existing." He also intended "to have these attached to the back of the badge." All workers in the highly secret project had to wear and display security badges. The union of film and badge, he noted, increased "the probability of the film and the individual being together."[16] It has remained a common practice ever since.

Wollan's plans took unexpectedly long to mature. The key problem was that film response varied with energy of X- or gamma radiation. Lower-energy rays produced more ions in emulsions than higher-energy rays. The same number of roentgens, in other words, at lower energies produced darker developed film. The answer was to use filters to flatten the response curve. The right choice of film and filter could give fairly even results over a useful range of energies. Finding the right mix took time. A year of research on films, filter materials, and filter thicknesses preceded the decision on an acceptable design. The choice was a cadmium filter 1 millimeter thick packaged with two types of film. One of the films carried a very sensitive emulsion which allowed darkness to be measured with reasonable reliability over a range from 0.03 to 3 roentgens. The second covered 1 to 20 roentgens. Only late in 1943 did film begin to join pocket chambers in tracking the X- and gamma-ray exposures of project workers.[17]

At the outset the most urgent question facing the Health Division was the most basic one from a safety standpoint. To how much radiation might workers safely be exposed? In one sense, setting protection standards was easy. The widely accepted prewar standards were ready to hand. Confronted with work well under way and workers already at risk, Wollan stated the division's clear choice. The radiation to which workers were exposed should be "less than what is agreed on as being safe (at present 0.1 r per day)."[18] In another sense, however, the problem was not so easy to resolve. What was agreed to be safe in the fall of 1942 could be regarded as only a stopgap. "Established tolerance doses" for X and gamma rays when the project began, Stone later explained, "rested on rather poor experimental evidence."[19]

Unreliable though their basis might be, X-ray and gamma-ray standards existed; the Health Division at least had a body of accepted usage for coping with the danger. For other hazards, even that was lacking. Little or nothing was known about tolerance for neutrons or for alpha and beta particles. For internal emitters—active substances swallowed, inhaled, or otherwise taken into the body and retained—the problem was more complex. The well-known hazards of radium and radon led to the first published standards in 1941. Less was known about radiophosphorus and radiostrontium, nothing at all about tolerance for other active nuclides. A tolerable burden of some active substance was the amount lodged in the body which could be borne without seeming harm. Measuring body burdens in living subjects was tricky, and then it provided only an after-the-fact notice that something had gone wrong. The best measure of body burden, in any event, revealed nothing about intake when metabolic pathways remained question marks. Intake, not burden, was the crucial safety problem, but no one knew much about the relation of amounts taken in to amounts retained. As intake was almost always inadvertent, measured data scarcely existed. The very meaning of tolerance dose for internal emitters was unclear. Repeatedly Stone insisted that the only safe practice was to try to avoid any intake at all.[20]

Enough was known in 1942 to provide, in Cantril's phrase, "base levels on which to build."[21] That was vital. Time pressed, and plans to produce plutonium had to proceed from "calculation of hazards based on such knowledge as already existed," Stone recalled. Yet no one questioned the need, he added, to check the assumed biological effects "by experimentation as rapidly as possible."[22] Other steps also could be taken. One seemed merely semantic. Although "tolerance dose" remained in use well into the postwar era, the preferred term changed during World War II. The new usage, "maximum permissible exposure," had begun to challenge the older term after the mid-1930s. In one sense, changing new words for old meant little. Numbers at first stayed the same. Maximum permissible gamma exposure was the same 0.1 roentgen per day the tolerance dose had been. In other ways, however, the shift was crucial. As the words themselves suggest, the new term connoted a ceiling, an upper limit to exposure, in contrast with the looser notion of tolerability implied by the term it replaced.[23]

At the outset, no new findings supported the change. It was rather a response to the clouded issue of radiation effects on living systems. Biological intuition and medical experience, in fact, suggested that persons exposed below some level would suffer no lasting damage. Given the large number of workers and the huge scope of the project, however, Stone and most of his colleagues preferred the prudent course. They acted almost as if they believed any exposure could be harmful, as if no threshold existed, as if any body burden might cause damage. They hoped research would provide better data than they had in 1942, but the project was too urgent to wait upon final results. Time enough later to raise the limits if research findings permitted. This course undoubtedly kept worker exposures well below what they might otherwise have been.

HEALTH DIVISION RESEARCH

Much Health Division research was more applied than basic: testing and proving the tools and methods required to meet the project's immediate health and safety needs. Typical was health physics work on film badges and pocket ionization chambers. But research was not limited to finding ways to measure individual doses. Instruments were needed to detect and monitor the whole range of particles and rays produced in nuclear reactions. The problems were challenging. Each radionuclide had its own characteristic decay mode. Each kind of radiation presented its own problems and hazards. Each workplace imposed its own special demands. Principles were all known before the war, but few had been reduced to the kind of rugged, standard, and reasonably easy-to-use instruments the project needed. The hard and complex task of turning theory into working tools lasted throughout the war and after.[24]

This effort was crucial to the entire project. Instruments were not only designed and tested at the Met Lab, but in many instances they were also produced in large numbers for use throughout the project.

They also served many purposes beyond health and safety. Health physics, in fact, worked closely with William P. Jesse's instrument group in the Met Lab's Physics Division. Jesse had received his Ph.D. in physics from Yale in 1924 and joined the University of Chicago a decade later. His research centered on X-ray crystallography, cosmic rays, and the ionization of gases. Like others engaged in such research, he learned a great deal about making instruments. He was among the first members of the Chicago faculty tapped for the Metallurgical Project.[25]

Ionization of gases, Jesse's specialty, was by far the most common basis for instruments. Nuclear radiation can be detected only through some effect of its interaction with matter. Of such effects, ions produced in gas are often the most useful. A device like the pocket chamber described above works electrostatically; the ions neutralize the charge. Connecting a power supply—normally a battery—attracts ions to the electrodes and allows the ion charges to flow in a circuit. This current flow can be precisely measured even at very low levels. This process in turn provides a measure of the radiation intensity that caused the ionization. Ionization chambers were one of the two main classes of instruments used at the Met Lab. They had been widely employed before the war in X-ray work. Commercial models were on the market and could be adapted without too much trouble to measure gamma rays, and they could also serve for beta radiation.[26]

The second class of instruments comprised proportional and Geiger-Müller counters. At voltages higher than those used by ionization chambers, freed electrons acquire enough energy to create new ions. The effect is called gas amplification. Over a certain voltage range, it increases current flow in proportion to initial ionization events. Devices working in this range are called proportional counters. At still higher voltages, proportionality fails. A single ionizing event may trigger an avalanche of electrons throughout the detector gas and produce a large electrical pulse. This is the working region of Geiger-Müller counters, named for the two German scientists who perfected the first such device in 1928. Proportional counters were, for the most part, still hand-made laboratory devices when the war began.[27] The Manhattan Project changed that because their major use was alpha detection, a special project concern. Plutonium is chiefly an alpha emitter. Massive and highly charged, alpha particles present little threat outside the body; even a sheet of paper will block them. But plutonium, like radium, proved to be a bone seeker. It might produce the same "horrible results . . . [as] radium salts taken into the body."[28] Alpha counters became the subject of intense work throughout the war. The result was an array of devices to check workers and to survey workplaces: hand, face, and foot counters, air samplers, and survey meters with nicknames like Pluto, Sneezy, Poppy, Zeus, and Juno.

Neutrons presented the most difficult problem for the instrument makers. Electrically neutral, their effects on matter differed widely from those of charged particles. From a safety standpoint, the hazard also depended closely on their speed and energy, which vary over a wide range. Monitoring high-speed neutrons in the presence of gamma rays presented special problems. One answer was a device called Chang and Eng, named after the famous Siamese twins because of its twin chambers; the first chamber measured the combined neutron-gamma effect, the second, gamma only. The current from the gamma-only chamber subtracted from the other chamber's current gave the neutron reading. Slow (low-energy) neutrons required a different technique, based on their reaction with boron to form lithium and alpha particles. Filled with a gaseous boron compound or lined with boron, alpha counters could thus be adapted to count neutrons.[29]

In a sense, instruments themselves might define the hazards. The Health Division adopted a tolerance dose for neutrons of 0.01 neutron unit per day. The unit was "that quantity of fast neutron radiations which will produce a reading of one roentgen on a Victoreen condenser r meter when using the 100 r chamber."[30] Exposure levels were often expressed as instrument readings. Thus the maximum permitted alpha level on work surfaces became 50 units on Pluto, the first alpha survey meter produced at the Met Lab. Five units on an alpha hand counter defined the limit for hands, 50 counts per minute on another device, that for faces. In the health physics section, research fell chiefly to Herbert Parker's protection measurements group. Instruments were not the sole concern. Parker also faced the old problem of finding satisfactory units to relate the physics of ionization to biological effects.[31]

The roentgen was defined only for X and gamma rays, but the project faced, Parker noted, the "practical problem of adding the doses received by a large group of workers from quantum radiation, alpha, beta, and neutron radiation."[32] The answer was a common unit. The key choice was to base it on energy absorbed rather than on ions produced. The "rep" (roentgen equivalent physical) measured dose as energy absorbed per unit mass (ergs per gram) equivalent at a point in the body to exposure in roentgens. Since biological effects varied with kind of ray, Parker derived a second unit that included a biological factor. Termed RBE (Relative Biological Effectiveness), it was found by experiment for each kind of radiation. The measure of biological dose was then the product of rep times RBE. Parker called it the "rem" (roentgen equivalent mammal or man). "Roentgen equivalent biological" might have made better logic; he rejected that term when he learned that hearers might confuse rep and reb.[33] Parker completed the new system in early 1944. Stone proposed it for the project at a Met Lab meeting on 7 March.[34] Simple, thoughtful, and convenient, Parker's system won many users during the war. Security prevented its public dissemination until 1948, but not until the late 1950s did rem, the dose unit, fully displace roentgen, the exposure unit, as the basic measure in radiological safety.[35]

The Health Division's medical section provided health services, but like the health physics section it conducted research as well. Clinical needs directed research chiefly in response to the project's direct health and safety demands. Leon Jacobson's blood studies and Albert Tannenbaum's toxicological work were typical. The research program also included other contracted studies on human subjects. Participating were Memorial Hospital in New York, the Chicago Tumor Institute, and the University of California Hospital in San Francisco. The subjects were patients not expected to survive more than a year or two. They received massive single, repeated, or continuous X-ray doses.[36] These studies, as Cantril explained,

> will give some useful information which we do not have on the effects of whole body irradiation. They may permit an appraisal of the effects after . . . at most 1–2 years. . . . They also serve to judge the magnitude of an exposure to which a man could be subjected in an emergency and the length of time before restoration (as well as can be judged from present laboratory methods).[37]

Researchers sometimes used themselves as guinea pigs; in one study of the effects of external radiation on white blood cell counts, four volunteers received up to 50 roentgens whole-body X-irradiation.[38]

Other research was pursued outside the Chicago laboratories. The Health Division sponsored or supported such efforts from the outset. Some of these emerged as major programs in their own right. At the National Cancer Institute in Washington, Egon Lorenz and his colleagues began a small-scale study in the spring of 1941. They exposed mice and other animals to relatively low levels of gamma rays over extended periods of time. Shortly after he joined the Health Division, Stone visited Lorenz. Cooperation between the two groups followed. Met Lab support allowed Lorenz and his co-workers to expand their studies with a much more elaborate experimental facility. Early in 1943 they could begin a new series of experiments using far larger numbers of animals.[39]

The project at the University of California in Berkeley also had pre-war roots. Joseph G. Hamilton had worked with Stone on studies of cyclotron-produced active nuclides, later found to be fission products as well. After Stone went to Chicago, Hamilton pursued the work under Health Division auspices. Such studies were badly needed, as Hamilton observed in his first report: "Of the 17 or more elements . . . produced by uranium fission, only a few have been studied with regard to their biological disposition."[40] Little was known about the fate of fission products once they entered an animal's body. One question had special bearing on tolerance. Did other radioelements share iodine's well-known trait of concentrating in a single organ or tissue? Biochemistry ignored radioactivity and thyroids collected all isotopes of iodine indiscriminately. Many fission products, however, were isotopes of rare elements with unknown biochemistry. What impact an active nuclide might have remained guesswork. Using micrograms of radioelements from the Berkeley cyclotron, Hamilton began tracing metabolic pathways of the products and their compounds.[41]

The line between applied and basic research often blurred. As Stone observed, "the clinical study of the personnel is one vast experiment. Never before has so large a collection of individuals been exposed to so much irradiation."[42] Kenneth Cole's experimental biology section, however, was clearly intended to focus the division's basic research. Stone was not the only one who believed that "beneath all observable effects was the mechanism by which radiations, no matter what their origin, caused the changes in biological material. If this mechanism could have been discovered, many of the problems would have been simpler."[43] Whether a full-scale basic research program could have been designed in 1942 must remain an open question. The dearth of sure knowledge may have been too large a handicap.

In any event, wartime pressures precluded such an approach. As Cantril and Cole noted early, "work directly connected with immediate problems of the Project" came first.[44] Xenon was an example. A radioisotope of that noble gas was known to be a fission product. It might be expected to diffuse into the air around a working pile. Although chemically inert, the gas might still build up in workers' lungs or tissues. No one knew. Cole tested animals to learn what the effects of breathing xenon were; Wollan computed the amounts likely to be inhaled. That fear was soon laid to rest, but others followed.[45]

"Immediate problems" persisted. That was one reason for such complaints about being "hampered by lack of materials, space and personnel" as Stone made in May 1943.[46] No research program ever commands all the resources its workers might desire. The researchers sustained a high level of enthusiasm and esprit de corps throughout the war, despite sometimes trying conditions. Part of Cole's section, for example, worked in a converted brewery stable near the Chicago campus. Typical of many former Met Lab researchers, one worker in that stable recalls the "tremendous enthusiasm [of] a very close knit group [which had] a kind of spirit I had never seen anyplace."[47] Many questions were never fully answered; some were never even asked. Yet much was learned about the action of radiation on cells, tissues, and organs, about the long-term and delayed effects of radiation on living systems, and about a host of related problems. In the long run the wartime work far exceeded in scope and value any that had gone before.[48]

THE UNIVERSITY AND THE ARMY

Not quite four months after the Health Division's birth, the first pile began running. Long expected, it still transformed the division's work. The only division member on the scene was Carl C. Gamertsfelder. Fresh from the University of Missouri with a Ph.D. in physics, he joined the project as a health physicist. He recalls struggling to measure the pile's radiation with an electroscope designed for cosmic-ray research; it had to be read through a built-in microscope.[49] As the pile ran at higher power levels over the next days, Stone later reported, meters showed that "no one connected with the operation could receive a tolerance dose of radiation." Even had it run without pause at unrealistically high power, "no one in the vicinity would have been injured."[50] And that included people living nearby. Compton was quite certain that gamma radiation from the pile was not likely to cause nausea or death in any campus neighbor.[51]

Now, however, the context altered. A series of technical choices and organizational changes transformed the project between October 1942 and February 1943. Essentially an academic research effort in the fall, it became by spring a growing army-directed industrial enterprise. Early plans to complete pile research on campus and then to produce plutonium in a new plant sited at nearby Argonne were scrapped. Pile research was reslated for Argonne. Partly to shift operating hazards from a major population center to a more remote locale, the plant would go to a still newer site in eastern Tennessee near Clinton, 25 miles from Knoxville. It became the Clinton Engineer Works, code-named Site X (and renamed Oak Ridge only after the war). As even the much larger X-10 area seemed too small before 1942 ended, still a third site, near Hanford in eastern Washington, was selected. The X-10 plant—piles to irradiate uranium and a chemical processing facility to extract plutonium from the fission products—became the so-called semiworks. It was designed as a pilot plant for the full-scale effort at Hanford Engineer Works, itself code-named Site W.[52]

The University of Chicago, after some soul-searching, agreed to direct the semiworks. Five hundred miles from campus, the job also fell well ouside normal academic pursuits. Circumstances were not normal, of course; the Met Lab clearly had the experts best suited for the task. The Chicago project, in fact, was growing more complex. Neither uniform nor systematic, the process persisted throughout the war. Initially, Plutonium Project and Metallurgical Laboratory had been all but synonymous, but as the project evolved it began spinning off other independent laboratories. From a small research contract with Iowa State College on methods to produce uranium metal grew the Ames Laboratory. Argonne Laboratory started life as a Met Lab division, then became independent in May 1944. Clinton Laboratories at Site X, created by the Met Lab to run the semiworks, became the Oak Ridge National Laboratory after the war.[53]

I. E. du Pont de Nemours and Company of Wilmington, Delaware, joined the project in 1942. The company, like the university, had qualms, though for different reasons. It still smarted from the "Merchants of Death" tag it received in the 1930s, a product of its role as a major munitions supplier in World War I. Yet, again like Chicago, Du Pont seemed to be most highly qualified for the job. It accepted army contracts to build the semiworks and both to build and to run the Hanford plant. Stipulating payment limited to costs, however, the company refused any profit from its work on the Manhattan Project. It also insisted on terminating its part in the project as soon as the war itself ended. Once Du Pont had decided to take the job, however, all trace of reluctance vanished. Its special expertise in large-scale industrial engineering proved as vital to the project as planners had expected.[54]

In December 1942 President Roosevelt approved a full-fledged program to produce bombs. The project had now grown beyond its current sponsor, the Office of Scientific Research and Development. As 1943 opened, building was under way in Tennessee. Work was about to begin at Hanford and at still another site, the weapons research laboratory in Los Alamos, New Mexico, where the metal would become bombs. Since the summer before, all construction was united under the Manhattan Engineer District, U.S. Army Corps of Engineers. Headquartered in New York, the MED was commanded by Colonel James C. Marshall. Veteran engineer officer and newly promoted Brigadier General Leslie R. Groves took charge of the bomb project in September 1942. His command included Marshall's MED, the research projects around the country, and much else besides. Over the next several months OSRD transferred all its contracts to the War Department. The pace of the work sharply increased.[55]

The Health Division was formed in Chicago at very nearly the same time the army began to assume control of the Metallurgical Project. For the first year or so, the change had little impact. The chief contact between the division and the army was through the person of Hymer L. Friedell. After medical training, Friedell had taken a degree in radiology from the University of Minnesota in 1940. As a fellow of the National Cancer Institute he did further work at the Chicago Tumor Institute, where he knew Wollan. Then he worked with Stone and Hamilton at the Crocker Laboratory in Berkeley. In August 1942 he enlisted in the Medical Corps. Friedell was the first medical officer assigned to the MED. He went to Chicago as informal liaison between the MED and Stone's group.[56]

At that time the Health Division also included a military section. It had two chief concerns: protecting troops in the field against radiation and studying the use of fission products as weapons. Both concerns stemmed from fears of Nazi nuclear research, a very real threat in the minds of Allied planners during the early years of the war. The immediate problem was not a bomb, which they knew would take years to produce. Rather it was the direct use of fission products as "a particularly vicious form of poison gas."[57] In other words, the same threat that demanded a health division to protect workers also suggested a military opportunity. A weapon might be derived from the highly active products of a chain-reacting pile.[58]

The first warning sounded in May 1941. Its source was a committee chaired by the same Arthur Compton later chosen to direct the Metallurgical Project. The best guess then was that a weapon using pile products might be ready twelve months after a first chain reaction.[59] A report issued in December warned: "The fission products produced in one day's run of a 100,000 kw [kilowatt] chain-reacting pile might be sufficient to make a large area uninhabitable."[60] Compton feared that could happen as early as 1943.[61] After taking charge of the Chicago project, he assigned J. C. Stearns to "a program designed to build up a defense against this form of attack."[62] Darol K. Froman was asked to assist Stearns with instrument design.[63] Stearns's first concern was the prospect of German air raids dropping radioactive poisons on American cities. He drafted a report on countermeasures.[64] He also asked experts around the country for help in forming crews to provide warning against such attack.[65] The Met Lab recruited its own team of experts on 24-hour call to respond should the threat materialize.[66] "We had to let people know where we were" at all times, one recalls. "We had emergency kits, ready to move on very, very short notice. I mean the quality of concern with respect to the German effort was high."[67]

In May 1943 Compton appointed Wollan to chair a special committee on the military potential of fission products. Its findings furnished the basis for a National Defense Research Council report to the army. It concluded that a pile could indeed provide fission products in amounts large enough to pose a real threat. Output might reach the equivalent of a ton of radium per week. Spread over 1 or 2 square miles, it could disable a major fraction of the population and force complete evacuation of the affected region.[68] Defense against such a threat was the chief concern, but offensive uses were studied, too, if not strongly pursued.[69]

Strategic bombing was not the only perceived danger. Tactical implications also concerned Stearns, who warned "that when our forces invade the Axis territory we should be adequately prepared to meet such eventualities."[70] This warning was heeded. "Peppermint" was the code name for a 1944 plan to prepare Allied forces in the invasion of Europe against German use of radioactive poison. General Eisenhower and his staff received a briefing on radiation hazards before D day. The Met Lab and other army contractors provided eleven survey meters and a Geiger counter, as well as 1,500 film packets, for shipment to England in spring 1944. They were to be deployed upon any report of unexplained film fogging in the field or of "epidemic disease of unknown etiology."[71] Another 150 instruments and 1,500 film packets waited in the United States for prompt air shipment to Europe should events dictate. Here the authorities were merely being cautious, for intelligence from several sources had by then allayed most concerns about the German program. As the planners had come to suspect, German fission products did not exist.[72]

RESEARCH UNDER FIRE

Little of the Health Division's work had so direct a bearing on military concerns. The division's role under army aegis, in fact, was not at all clear. What was clear in the spring of 1943 was that it needed sharper definition. The project was growing swiftly. Operations soon to begin at the semiworks in Tennessee would require a health and safety staff in short order. So would the Hanford plant in Washington not too much later when plutonium production got under way. As the army prepared to assume full control of the project, hard questions were being asked about who was to do what among the university, the army, and the industrial contractors, especially Du Pont. Such questions were not confined to health and safety, but neither were matters of health and safty held to be minor. Groves and Marshall conferred with Compton and his chief lieutenants early in April 1943.[73] Stone was among those trying "to clarify the health responsibilities of the Army as opposed to those of the University of Chicago Project."[74]

They agreed on a number of points. The Health Division would be wholly in charge of health and safety at Site X. That responsibility included the research required to protect the public as well as the workers. The division would continue to inspect and suggest safeguards for groups working under Met Lab contracts. It would also advise related projects. "In particular," Compton noted, the Health Division "shall advise and cooperate with Du Pont in preparing to meet the hazards associated with their operations related to this project."[75] Du Pont could be expected to handle normal industrial hazards, Compton explained, but Stone's division must shoulder the full burden for "hazards arising from radiation and radioactive materials."[76] Formal transfer of the project to the army came on 1 May 1943. The new army contract required the university to create and maintain the proper health and safety measures. It also approved "such studies . . . with respect to biological and medical problems related to . . . the development or use of the process as may be necessary or advisable in the judgement of the contractor."[77]

Friedell was still serving as the chief link between the Health Division and the Manhattan Engineer District. He also remained the MED's only assigned medical officer. Since March 1943, however, the district had also been receiving advice from Stafford L. Warren. A 1922 medical graduate of the University of California, Warren had been at the University of Rochester since 1926 and professor of radiology since 1933. Rochester was also the home of Eastman Kodak, which provided his link to the Manhattan Project. The company's operating arm, Tennessee Eastman Corporation, received the army contract for the electromagnetic separation plant at Site X. In March 1943 an Eastman Kodak vice president arranged the secret meeting with Groves and Marshall which brought Warren into the project. Although still a civilian, his role in MED health and safety planning became central. After a change in command during the summer of 1943, it also became formal.[78]

In July 1943 Kenneth D. Nichols replaced Marshall in command of MED. A 1929 West Point graduate, Nichols also held a doctorate in engineering from the University of Iowa. One of his first moves transferred MED headquarters from Manhattan to Site X in Tennessee. He also formed a medical office and named Friedell its executive officer. Defining its mission in broad terms, Nichols endorsed as the first task "the conduct of, arrangement of, supervision of, or liaison with, such research work as is considered necessary for carrying out the functions of the Medical Section."[79] Nichols also charged the medical office to decide what health hazards existed and institute proper measures against them throughout the Manhattan District; to instruct contractor and MED officers on the need for health safeguards and the proper use of medical services; and to inspect all programs to assure that ordered measures were obeyed. Warren helped plan the new office, in fact drafting the statement of mission issued over Nichols's name. In November he accepted a Medical Corps commission as colonel to head the medical office. At the same time he was named chief medical advisor to Groves. The younger and less experienced Friedell remained as Warren's deputy.[80]

The Groves-Compton talks, the Met Lab's contract with the army, the charter of the MED medical office—all seemed to confirm that army and university agreed on the need for research to safeguard workers against project hazards. But the matter was not so simple. Studies of acute radiation effects, of fission-product metabolism, or of other questions clearly related to running large-scale plutonium plants were never an issue. In fact, Warren augmented such efforts with a contract to the University of Rochester for a major research program in uranium toxicology.[81] The problem lay elsewhere. Lengthy studies of the effects of chronic low-level exposure, basic research on the action of radiation in living systems—such efforts were not so easy to defend in the context of a wartime project.[82] As Compton delicately remarked in his postwar memoir, "There was some difficulty in convincing the Army that this physiological work was of the importance implied by our rather extensive program."[83]

At Warren's request, Stone and others prepared a report on the Health Division's research program in April 1943. A fuller self-evaluation followed in May.[84] The second report, in effect, proposed "a marked expansion of the work of the Biological Sections of the Health Division."[85] The army approved the plan, but then delayed funding. In the end it never did provide the extra money. The nature and the scope of a proper wartime research program were no easy matters to resolve. Biological research was not the only aspect of the Metallurgical Project to become and remain a source of friction between the university and the army.[86]

In mid-1944 Stone and Warren again tried to resolve the issues. They agreed to focus research on the effects of acute rather than long-term or chronic exposure and to build no new facilities.[87] From Warren's viewpoint, the value of project research lay in quick, useful results. As he noted for the record when the issue arose once again a year later, results could not always be predicted. Nevertheless, programs unlikely to produce short-term findings ought not to begin, and research not soon applied should be dropped. Policy, he insisted, had always stressed worker safety through "adequate protection and precautions, thereby reducing the operating problems to a minimum." Research was not an end in itself. Rather it supported the proper concerns of the

Health Division: setting tolerance standards, monitoring work, and enforcing safety practices in accord with accepted industrial standards. The MED, in short, "is charged with conducting operations which will be valuable in the present war effort; . . . all accessory and auxiliary functions should be directed and oriented so that they are in direct support of this premise."[88]

These views were not fully shared in the Health Division. Stone responded with a detailed critique, stressing two points. Autonomy of his division's research effort was one. Not only begun well before MED entered the picture, it remained under constant self-review. The role of MED Medical Office was to help and advise, not to direct. The army-university contract, he pointed out, nowhere mentioned research supervision. "[H]ad such a statement been included the Associate Project Director for Health [Stone] and the great majority of the top men in the health program would not have remained." The second point mattered more. Stone denied "the direct winning of the war" as the crucial factor. When biological research began in 1942, no one knew how long the war might last. The university, in any event, had a "moral obligation to the personnel and the community" that must "extend far beyond the war period." The university's price for taking on the Plutonium Project included the army's agreement "to allow it to do the biological and medical research work it felt was" required to meet that "moral medical responsibility."[89]

Opposed viewpoints about the proper nature and scope of research may have echoed deeper conflicts. Stone strongly objected to Warren's claim that one purpose of research was "to 'strengthen the Government interests' from the medico-legal point of view." They would be amply protected, Stone believed, "if the health and welfare of the workers are the main objects."[90] Much the same thinking inspired his support for the shift from tolerance to maximum permissible exposure as the central tenet of radiation safety. Warren, in contrast, argued that normal industrial safety practices were good enough. Special standards should apply only upon proof of prompt, clear-cut biological changes or health threats.[91] Friedell was even more forceful. Setting permitted levels of plutonium too low, he warned, could backfire. Although plutonium was no worse a hazard than other common products of science or commerce, it was new and unknown. Standards too stringent posed "the danger of untoward psychological effects." Strong efforts, of course, should be made to control the hazards "to as low a level as is reasonably possible," but not at the expense of the project. Levels too low "might impede the production program." That he found "not compatible with the overall urgency of the Manhattan District operation."[92]

Friedell's viewpoint was neither callous nor unreasonable, nor, for that matter, is it outmoded. Controversy over proper radiation safety standards since then has often hinged on just such points as were raised during the war. The issue was never simple, nor were sides ever merely drawn between those in uniforms and those in lab coats. Cole and Jacobson, for instance, both recall how urgent the wartime program seemed. In that context they agreed with Friedell on the legitimacy of having workers take some risks. Such risks were, in any event, far smaller than those routinely faced by front-line soldiers.[93] "One of my standing complaints," Cole later observed, "was that M.D.'s did not allow for the fact that there was a war on and insisted upon peace time standards of safety."[94] Pragmatism, in part, prevailed. War held dangers that all might reasonably be expected to share, one doctor noted. Yet actual policy stressed strict controls intended to prevent "all the conditions associated with delayed injurious effects" from radiation.[95] How large a part basic research should play in forming such policies remained an open question.

OPERATIONAL HEALTH AND SAFETY

The place of basic research in a wartime project may have been an important long-term policy issue. It had little bearing on day-to-day efforts to ensure worker safety, which were growing as the Manhattan Project expanded. By summer 1943, Site X in Tennessee began to impose its demands for medical and health physics staff. Chicago was the major source of the needed experts. It also provided most of the training for those who came from elsewhere. The Met Lab devoted fully a third of its first training course for Du Pont employees to health and safety.[96]

The first ranking member of the Health Division to reach Site X was Cantril, who traded his title of medical section chief for medical director, Clinton Laboratories. Parker, his long-time colleague, soon joined him as chief of the new health physics section. Reorganization of health physics at Chicago was substantial. Wollan left the Health Division for the new General Physics Division, taking with him most of the instrument makers. Louis A. Pardue, a Yale physics Ph.D. and one of the

Met Lab's original group of health physicists, took over as group leader in the Health Division with a small staff; most of the rest went to Site X with Parker.[97] Potential hazards at the plutonium pilot plant loomed far more serious than any posed by the research piles near Chicago. Biological research still centered in Chicago, but Site X would also house a program. Howard J. Curtis, a biophysicist and former colleague of Cole's at Columbia, was recruited to direct the effort.[98]

Du Pont only built, while Chicago ran, the semiworks at Site X. The company would both build and run the full-scale production plant at Site W in Washington. Du Pont had emerged during the 1930s as a corporate leader in the still novel field of industrial safety, highly respected for efforts to protect workers in the notoriously risky manufacture of explosives. Those concerns carried over to its work in the Manhattan Project. The company strongly supported safety measures, even at the expense of delays to what was, after all, a wartime crash project. Critics were not lacking. Leo Szilard, for one, claimed to voice a widespread view when he referred to Du Pont's "irrational [and] exaggerated considerations of operational safety."[99] As Parker notes, however, the company made " 'Safety First' real, not just a trite slogan."[100] Yet Du Pont began no more prepared than any other producer to cope with the special hazard of radiation. Two doctors from its industrial medicine division, W. Daggett Norwood and Philip A. Fuqua, went to Chicago for special training. Both then transferred to Site X to bolster the medical staff there and to receive further training. Norwood was slated to become medical director at Hanford. Fuqua would be his second-in-command and Parker his chief health physicist. Between the summer of 1943 and the spring of 1944, however, new fears about plutonium forced a change in plans.[101]

Although everyone expected plutonium to be a major hazard, it had received little study before 1944. One reason was that so little plutonium existed—a total of some 2 cyclotron-produced milligrams—until the fall of 1943. Other hazards also seemed more urgent: neutrons and gamma rays from the piles, the fission products from which plutonium must be extracted. Then the semiworks began to produce milligrams by the end of 1943, grams within a few more months. At the same time, chemists now facing the prospect of working with plutonium started voicing their concerns. Efforts to define the hazard began at Berkeley in February 1944 with studies of the metal's toxic effects. Work would be well under way, however, long before any final results were achieved. The project set a provisional tolerance level of 5 micrograms (or about a third of a microcurie) retained in the body; it derived from the radium standard, adjusted to plutonium's lesser activity. The half-life of radium 226 is 1600 years, that of plutonium 239, 24,400 years. Work began at once to alter techniques and workplaces at Chicago and the Clinton Laboratories to meet the standard.[102]

These changes also meant changes at Hanford. The experienced Cantril joined the Du Pont team at Site W as head of the industrial medical department. His express charge was to deal "with the health and protection of the workers from any hazards peculiar to the operations."[103] To replace Cantril at Site X, the project found John E. Wirth, the man whom Cantril had succeeded as director of the Tumor Institute in Seattle. Wirth's longtime associate, health physicist John E. Rose, also joined the project. Wirth and Rose had expected to join the Hanford team after training at Site X. In the event, Wirth remained as medical director of Clinton Laboratories, while Rose became chief health physicist at Argonne. Karl Z. Morgan, an early recruit to Chicago health physics, took over at Site X when Parker left for Hanford.[104]

Despite the fears, the health and safety program at Site X proved strikingly mundane. As Wirth remarked, more effort went into handling "the general industrial work than it took to cope with the special hazard problem."[105] Industrial hygiene, routine first aid, and public health measures demanded far more concern at the remote site than radiation hazards.[106] The biggest problem may have been workers' lack of knowledge. In contrast with Chicago, both the Tennessee and Washington sites employed large numbers of craft workers, factory hands, and laborers. Knowing nothing of the "special hazard," some workers grew anxious about blatant but seldom explained measures to protect them. Staff members often felt obliged to assure workers that their ailments had nothing to do with "those gremlins up on the hill."[107]

At the heart of radiation safety were strictly enforced safeguards against worker exposure. Measures at Site X followed the Chicago pattern based on well-known industrial practices for handling toxic substances. Included were techniques for remote handling of active material; the use of special clothing, gloves, shoes, and masks; and the design of laundry and decontamination procedures. Access to so-called hot areas—places where the work involved radioactivity—was closely controlled. Both workers and workplaces were subject to constant monitoring.[108] The Health Division took pains to stress "the desirability of as low an exposure as possible and the need for the prevention of exposures in excess of 100 mr [milliroentgens] per day," Wirth recalled. Exposed workers' records were checked, their supervisors required to explain. Repeated exposures were followed by talks with the workers, study of their jobs, and plans to avoid future exposures. Workers could scarcely avoid being impressed by such efforts. Some groups, Wirth reported, set their own limit at 50 milliroentgens to prevent any of their number from reaching 100 milliroentgens per day.[109]

Clinton Laboratories mounted a large-scale screening program. Like Chicago's, its results were meager. The effort involved, Wirth wryly commented, "more search than discovery; the Project was fraught with tremendous potentialities for overexposure" but there were none "of serious import."[110] The first effects of exposure were then believed to be on the blood-forming organs. All employees working in hot areas, about half the work force, received routine monthly blood counts. The other half were checked on a three-month schedule. Routine urinalyses followed the same pattern. Neither blood nor urine showed signs consistently linked to the relatively low-level exposure workers received. Such results did nothing to lessen the effort. The absence of any certain signs of damage might simply reflect the limited value of known tests, as Cantril and Parker had pointed out during the war.[111] It remained true later, when Wirth noted the "still urgent need for a means of detecting subclinical damage so that the unexpected appearance of dangerous changes months or years after exposure may be prevented."[112] In the absence of such means, the only recourse was constant caution.

Health and safety procedures followed much the same pattern at Hanford. The bulk of the work was routine, while concern focused on "protecting personnel and the public from overexposure to radiation in any of the forms encountered."[113] As elsewhere, such protection meant strict controls over the workplace and frequent tests of the workers. Worker training and indoctrination were hampered, at Hanford as elsewhere, by tight security. Many workers simply lacked the clearance to be informed of the nature of the hazards their work entailed. Like those in Tennessee, they needed reassurance, especially after the bombs had been dropped on Japan and they became aware of what they had been doing.[114] Ironically, as Parker noted many years later, the statement he and Cantril drafted to reassure the workers was defeated by security; it remained classified and thus inaccessible to those it addressed for two years after the war.[115]

Training and indoctrination had become a growing part of the Health Division plan. Constant reminders of the hazards and directions on how to cope with them helped workers protect themselves. They also helped avoid stalling the project.[116] Contamination could remove an entire building from use. "What widespread contamination can be caused by a minute quantity of hot material," exclaimed an amazed Wirth. "No less amazing is the ease with which it seems to get about, as though it were a living creature trying to spread itself anywhere."[117] Contamination also meant losses, and scarce resources made even small losses from careless handling a real problem. Savings from better practices, in Stafford Warren's judgment, "amply repaid the cost of the changes necessary to tighten up the program and carry out the safety and health precautions."[118]

At least through 1944, Manhattan Project safety concerns centered on making and handling plutonium, but they were not the only concerns. The project was widespread. That meant, among other things, extensive shipping between sites where special work was being performed.[119] As Groves remarked, trying to convey some idea of the problem's scope, "Fifteen different contractors perform[ed] hazardous work in connection with experimentation, development of processes, and actual operations of one phase of metal-working alone."[120] Many of the places where such work was done were small. One major task of the MED medical section was trying to ensure that each plant conformed to safety standards. This required army and Health Division experts to inspect work throughout the country.[121] Their visits were viewed, Warren recalled, with "apprehension and at first with open hostility." In time friction lessened; contractors learned that required safety measures "were really quite simple and chiefly based on good housekeeping."[122] The effort, however, was no great success. Work was often performed under conditions that would not have been allowed by the Met Lab or Du Pont.[123]

Although workers might face more serious hazards, public safety was also a matter of concern from the outset. It ranked with secrecy as a reason for the remote siting of the major production plants. In both Tennessee and Washington, large controlled areas served as buffer zones between workplaces and the outside world. Monitors in large numbers, many of them army enlisted men, were trained to use survey meters and assigned to patrol the empty areas. Air and water samples were systematically collected and tested for signs of radioactive contamination. Environmental concerns were reflected in research as well. Clinton Laboratories probed effects on animals exposed to pile waste products or stack emissions. An MED contract with the University of Washington began a major study of radiation effects on fish. The Hanford plant used massive amounts of Columbia River water for cooling. What effect that might have on the river's major salmon fishery prompted support for the program.[124]

Because making plutonium posed the most difficult problems, the work of the Chicago Health Division and its spin-offs at Oak Ridge and Hanford became the wellspring of Manhattan Project radiological safety. Despite the vastly expanded hazards of the wartime project, safety remained firmly rooted in prewar practice. This dependence on earlier standards caused no surprise, as the men in charge were themselves the product of prewar experience. Whether or not the new scale meant a change in kind as well as in degree of hazard is an open question. Old or new, the health and safety program of the Metallurgical Project was a success by the standards of its time and place. That success, as Robert Stone observed fairly enough, "is attested amply by the fact that no one lost his life because of any of the hazards peculiar to the Project." Under often harsh conditions, among so many thousands of workers, not every injury could be fully explained. Yet, Stone added, "only slight evidence of injury to any of the employees was ever detected, and, even in the cases in which injuries seemed to have occurred, the connection with Project activities was not absolutely clear."[125]

Health Division research was intended to make such links clearer, if they existed. The division began and defended a research program just because so much was unknown about the hazards and their effects. The question of what range and depth of research to pursue posed issues more pragmatic than theoretical. Disagreement within the division, and between the division and the army, reflected differing judgments of what might be practical in the short-term wartime setting, rather than deeply opposed views about the long-term value of further research. The men who took charge at Chicago in 1942 were well aware of their ignorance. Yet they had to make choices in the absence of perfect knowledge. The passage of time has brought questions about some of those choices. Further study has provided a larger base from which to assess the hazards. Perhaps more important, changing times have altered the social setting in which radiation safety is practiced.

But in the light of what was then known, in the setting that then existed, the Health Division held the risks to tolerable levels and laid the groundwork for much of what was to follow.

3

RADIATION SAFETY AT
LOS ALAMOS

THE LOS ALAMOS HEALTH GROUP

Louis H. Hempelmann, Jr., reached the secret weapons laboratory at Los Alamos early. The young doctor had been called to head radiological safety on the remote mesa in north central New Mexico. Like others chosen for the top-secret project, Hempelmann knew the right people. After medical training at Washington University in St. Louis, he interned in Boston. Then a fellowship brought him to California's Berkeley Radiation Laboratory in 1941. He studied radiobiology with John H. Lawrence and Robert Stone, future head of the Chicago Health Division. Also at Berkeley was J. Robert Oppenheimer, the physicist selected to direct Los Alamos research. When Oppenheimer decided early in 1943 that he needed someone for radiation safety at Site Y, he contacted Lawrence. Lawrence declined, instead suggesting Hempelmann, then in charge of Washington University's new cyclotron. Oppenheimer went to St. Louis and offered him the job. Not yet twenty-nine and less than five years out of medical school, Hempelmann suffered qualms, but he found the offer impossible to refuse.[1]

Before doing anything else, he caught a train to Chicago. He wanted to talk to Stone and other members of the Health Division about what to expect. They told him the hazards would be comparable to those arising from the use of supervoltage X-ray machines; no more than fifty or sixty people might be at risk, and the main task would be checking their blood counts.[2] This proved a fair preview for the first year of the job at Los Alamos. On his return, Hempelmann got in touch with an old friend and classmate, James F. Nolan, then studying at Memorial Hospital in New York. Peer pressure had induced him to apply for a commission in the Army Medical Corps. As the decision was still pending, Nolan welcomed the prospect of joining an important secret war project. He saw Oppenheimer in Washington and accepted a formal offer, still knowing little beyond the New Mexico location. After a stop in St. Louis, Nolan joined Hempelmann aboard a westbound train for a first look at their new station.[3]

They met Oppenheimer in Santa Fe. A frequent visitor to the region since 1929, Oppenheimer was on familiar ground. West of Santa Fe the land rose to the Pajarito Plateau, divided by narrow canyons into a series of mesas. Beyond, still higher, lifted the Jemez Mountains, while eastward towered the Sangre de Cristos. The rugged and beautiful setting had become something of a playground for the wealthy. Dude ranches dotted the landscape. Los Alamos itself, before the army claimed it, had housed the private Los Alamos Ranch School for Boys.[4] A few school buildings survived as the only grace notes amid the functionally drab army buildings rising on the site. "Plain, utilitarian, and quite ugly," remarked a later arrival, "but surrounded by some of the most spectacular scenery in America."[5]

Oppenheimer drove the two doctors to the site. The trip began on two-lane blacktop and ended thirty-five miles later after a heart-stopping ride up the gravel switchback to the mesa's summit. Construction was still in full swing. Oppenheimer and a few other scientists had moved to the area in mid-March, but they were quartered at nearby ranches. The only full-time inhabitant as yet was a single guard equipped with a sleeping bag. Site Y was an army base. The arrival of 250 troops, most of them military police, marked the formal start of work on 15 April 1943. Barbed-wire fences ringed the entire site. Military police checked passes at the gate and patrolled the fences. Within the compound, other guarded fences controlled access to the technical area and its research buildings.[6]

After their late March visit, Hempelmann and Nolan went home to fetch their families. When they returned a few weeks later, research at Site Y was getting under way. Hempelmann and Nolan divided the work between them. The Army Medical Corps approved Nolan's request for a commission. Appointed captain early in June, he took charge of the post hospital. His main task became normal medical care for those stationed at Site Y. Hempelmann, who remained a civilian, became Health Group leader. He concerned himself chiefly with health and safety in the laboratory proper, the technical area. In contrast with the divisional status accorded health at the Met Lab and elsewhere in the Manhattan Project, the Los Alamos Health Group remained a staff office under Oppenheimer throughout the war. Hempelmann's key charge was guarding the workers against the "special hazards."[7]

This task imposed no heavy demands during the first year. Theoretical work predominated. Particle accelerators and radioactive sources did raise some concern for stray neutrons and gamma rays and did define a small problem, a shortage of instruments noted by Hempelmann in his August 1943 health report. Laboratories could have used more ionization chambers; workers, more pocket chambers. Still, the number of workers was small, the risks were both reasonably well known and relatively minor; enough meters were on hand to pass around as needed without too much trouble. Hempelmann's main job was blood-counting. To handle it he needed only the help of a few technicians and a part-time secretary.[8] He did find himself growing concerned about the meaning of widely varied blood counts, but that proved less a problem than he had first thought. "The extent of variation in normal blood counts was not well known at this time," Hempelmann reported; "the Health Group became quite concerned over changes in the blood counts which, in retrospect, were within the limits of normal variation."[9] Instruments and blood counts seemed only minor problems.

In contrast with Chicago, Los Alamos planned no health-related research. Hempelmann expected the Health Division and other research groups to meet all his needs.[10] His only real concern was watching the workers. This entailed heeding "the suggestions of the Chicago Health Group," he explained: "frequent surveys of the laboratories to determine radiation intensities, daily record of exposure of each person, monthly blood counts (oftener when safe daily dose is exceeded), and semi-annual physical examination."[11] Hempelmann's duties left him time enough to double as post safety engineer and to help Nolan in the clinic. The two men never drew sharp lines between their respective assignments. Nolan also assisted Hempelmann; he later, in fact, became a full-fledged member of the Health Group. Retrospectively, Hempelmann could stress the relatively trouble-free nature of his first year at Site Y.[12]

As late as January 1944, hazards at Los Alamos differed little from those at any prewar laboratory using active sources and X-ray machines.[13] Then plutonium started to arrive. Health experts had, of course, foreseen the danger. Lack of plutonium, however, stymied research until the pilot plant in Tennessee began to produce. With the flow now starting, study of biological effects could begin in earnest. But other work could not wait. The chemists felt particularly severe pressure to devise techniques for separating and purifying the new metal. They grew alarmed about the "danger to personnel from inhaled or ingested plutonium."[14] Joseph W. Kennedy headed the Los Alamos Chemistry-Metallurgy Division. He came to Site Y from Berkeley, where he had been a colleague of Glenn Seaborg's in the 1941 discovery of plutonium.[15] Just as Seaborg raised the issue with Stone at Chicago, Kennedy asked Hempelmann for help. Throughout the project, only the imminent prospect of working with large amounts prompted a hard look at the practical problems of handling a substance little known but much feared.[16]

Even at this early date, fear ran deep in the ranks. Anxious chemists caught the ear of Cyril Stanley Smith, chief metallurgist and associate leader of the Chemistry-Metallurgy Division. Insurance worried them, and Smith fully agreed. The standard extra-hazard policy in force provided nothing, he noted, "for the real hazard involved in our work— radioactive poisoning from inhaled or ingested plutonium." Symptoms might take years to appear. Yet the policy covered no illness or disability that occurred more than ninety days after an accident or thirty days after termination. Smith called that "inhumane, unethical, and unfair."[17] Wartime secrecy made the problem doubly hard to solve. A partial answer emerged at length in a special arrangement between the army and the University of California. A secret million-dollar fund allowed the university to pay up to $10,000 to a worker (or his heirs) injured by one of the specified extra hazards.[18]

Relatively little certain knowledge of the hazards existed early in 1944. Plutonium seemed to behave much like radium in the body. It, too, tended to become fixed in bone. Its alpha emission, however, seemed to be fifty times smaller. The body might thus bear up to 100 micrograms of plutonium as against 2 of radium. The question of inhaled plutonium dust was harder to answer. Calculations could equate 1 microgram retained in the lung with 1.2 roentgens per day. That seemed well above any accepted limit, but very little was certain in this area. Standards for inhaled dust did not exist. As a first guess, the Manhattan Project adopted 5 micrograms as the tolerance dose for plutonium, fifty times the accepted standard for radium.[19]

The long, hard task of learning more began quickly. In February 1944

Oppenheimer wired the director of the Metallurgical Project to clear Hempelmann for a visit to Berkeley. There Joseph Hamilton's group under contract to Chicago studied the metabolism of fission products in rats. Oppenheimer wanted to make sure that the new research on plutonium metabolism met Los Alamos needs. "We are not equipped for biological experiments," he explained.[20] Hamilton had just been consigned 11 milligrams of plutonium, the first large batch provided for biological studies, and among the first for any research. Preliminary findings came promptly. The Met Lab's monthly progress report for February sketched the results. Plutonium was indeed a bone seeker and could be expected to cause radiumlike damage, including bone cancer. Plutonium was much less readily absorbed from the gut than radium, but it was also excreted much more slowly and retained much longer in the lung.[21]

Further studies confirmed these early findings. They were surprising. Plutonium seemed at least five times, perhaps even ten times, more toxic than implied by comparison of its alpha activity with that of radium. One key factor seemed the body's much slower rate of eliminating plutonium. More important was the way the two substances collected in bone. Although both were bone seekers, radium tended to spread throughout the bone volume. Plutonium, in contrast, collected mostly on bone surfaces, closer to the blood-forming marrow. The combined effects seemed almost to balance the hazards. Plutonium apparently posed very nearly the same risks as radium. The Manhattan Project lowered the tolerance dose for plutonium to 1 microgram. That change took fully a year, however, and even then the new level remained provisional.[22] Doubts persisted about the exact ratio between amounts entering the body and amounts retained. The war ended with the matter still unsettled. Value judgments as well as research findings shaped such decisions no less than the broader issues of biological research discussed in chapter 2.[23]

Plutonium may have been the most serious active hazard at Site Y, but it was not the only one. Uranium aroused little concern. Although used in large amounts, it was known to be relatively harmless. About polonium, health experts were not so sure. Its role, though small, became vital when bomb design was altered in mid-1944. Mixed with a light metal like beryllium, it provided neutrons to help initiate the plutonium chain reaction. Although an alpha emitter, it appeared to cause less damage than plutonium. Again in contrast with plutonium, it was not a bone seeker. Polonium was very quickly absorbed by the body but also quickly excreted. Other factors promoted the picture of a lesser hazard: smaller amounts used, simpler technical operations. The availability of a quick, easy test for polonium in urine, something achieved for plutonium only after strenuous efforts, also permitted routine checking of workers from the outset. Yet even at war's end much remained unknown about the way both polonium and plutonium behaved in the body and about the full extent of the hazard.[24]

SAFETY AND RESEARCH

Meanwhile, however, safety demanded prompter action. With other project doctors, Hempelmann visited a radium-dial firm in Boston for a firsthand look at hazard controls. His report guided Kennedy to set up three committees to begin work on the problems, one of which was assigned to devise guidelines for the Chicago instrument makers. The pressing need was for portable alpha counters and air samplers.[25] Once again Oppenheimer stressed how much Los Alamos relied on Chicago. Experience and resources put "your laboratory," he noted, "in a far better position than ours to undertake [this work]."[26] The second committee began working on designs for plutonium-handling equipment; the third, on setting rules for working with active substances. At the same time Kennedy formed the monitoring and decontamination group under Richard A. Popham. Trained as a botanist but diverted into other channels by the war, Popham's new job was to head the Chemistry-Metallurgy Division's "central office for the control of the special radioactive health hazard." The first task was to begin "regular monitoring for activity of the dust in all laboratories." Then, as soon as he could, Popham would enlist specialists "for clean-up of rooms in which the hazard has gotten out of hand."[27]

Knowledge about plutonium and proper means to detect it were both in short supply. Popham worked closely with Hempelmann to devise expedients. One was the so-called swipe method of rubbing a 1-square-inch piece of oiled filter paper across a working surface and then testing the paper in a fixed alpha counter. Recorded counts per minute served as a rough index of concentration. Both men would have preferred more direct survey methods to this crude makeshift, but lack of portable alpha counters limited options. Swipes also ignored the key problem: airborne active dust workers might breathe. Chicago work on air samplers had as yet produced no results. That meant a second stopgap, nose counts. This technique involved swabbing inside a worker's nostrils with filter paper, again measured in the fixed counter. Readings higher than 100 counts per minute were somewhat arbitrarily deemed to require action, such as warning the worker to improve his technique or perhaps shifting him to a plutonium-free job. Nasal swabs provided at best only a rough index of how much dust might have been inhaled. The real problem, however, remained too little knowledge about the ways plutonium behaved in the lungs and elsewhere in the body.[28]

William Jesse's Chicago instrument group was neither recalcitrant nor incompetent. Radiological safety had not demanded alpha monitoring before the war. Although an alpha source, radium also emitted much-easier-to-measure strong gamma rays. Fission products, too, tended toward strong gamma emission. As pile radiations and fission products seemed the most urgent problems during the project's first years, the instrument group focused its limited resources on these areas. Plutonium posed the first major hazard that could be detected, essentially, only through its alpha activity. Methods existed, but largely in the form of clumsy and touchy laboratory devices. Without commercial sources to draw on, the Chicago group had not only to design and develop, but also to produce, what was needed. Los Alamos stood last in line. Chicago itself, the pilot plant in Tennessee, the production plant at Hanford—all required instruments. Both more closely linked to Chicago than Site Y, Sites X and W might well make demands easier to hear. Their demands also came sooner. It was, after all, only when plutonium began to arrive from Tennessee and Hanford that Los Alamos called for help.[29]

Shortages, however, accounted for only part of the problem. Each site had special needs. Even when developed, the large, fixed alpha counters and air samplers suited for production plants seemed too clumsy for Los Alamos. At Site Y concern centered on laboratory monitoring which required, above all, good portable survey meters. The only one yet available early in 1944 was the Met Lab's Pluto, an ionization chamber adapted for alpha counting. From Los Alamos it looked far too insensitive. Pluto shortages thus only partly explained why Site Y monitors persisted with the swipe method. For all the slow and spotty results, swipes could detect much lower alpha levels.[30] Los Alamos electronics experts strove to fill the gap. Their first effort produced a new hand counter. Installed in Building D—where plutonium work centered—it was ten times more sensitive than its Chicago counterpart. The chief need, however, remained a portable survey meter, and by fall 1944 Los Alamos was working on its own design. Led by Richard J. Watts, a special group in the Electronics Division was assigned nearly full time to health-instrument development.[31]

Even the best of instruments, however, might not have mattered. Physical scientists tended to believe that biological damage from radiation could somehow be fully reversed, "that in due time a cure would emerge."[32] Biologists and doctors were less hopeful, although they devoted intensive study throughout the war years to a wide range of treatments.[33] Plutonium also was new; in some scientists it excited more fascination than fear. "Now that I am out of reach of Louis Hempelmann and the health physics group," said Frederic de Hoffman, "I must admit that I . . . handled some of the first quantities . . . in an uncoated form without any real feel for its highly poisonous nature."[34] Plutonium corroded in moist air. To prevent that, and also to keep workers from direct contact with the metal, chemists devised an impervious coating. Protecting a scarce and precious resource and safeguarding workers need not conflict, as this instance shows.[35] In other instances the balance was harder to strike.

The nature of the work compounded the problem. Building D was designed to protect the metal from contamination, but the real problem, as the project's official historian observed, became guarding workers from plutonium. For this purpose the building was ill suited. "As larger amounts of plutonium began to arrive, adequate decontamination became increasingly difficult."[36] The work proceeded under intense pressure. Frequently changed experimental setups and often makeshift apparatus made routine safety measures hard to apply. There was, indeed, no real basis for such routines. Even so common a standby as laboratory glassware might betray the unwary user; two major spillages of dissolved plutonium in August 1944 were blamed on glass weakened by prolonged alpha exposure.[37]

Safety practices clearly needed upgrading. In his health report for August 1944, Hempelmann announced a new approach adjusted "to suit each particular problem." Proper measures, he argued, "can only be developed after a thorough consideration of the problem by the Health Group and the individuals concerned."[38] Accordingly, the Health Group arranged a series of meetings with other groups during August and September. The goal was matching recommended safety techniques to each group's special tasks. Plutonium toxicology became the subject of Health Group lectures as another way of driving home the health hazards. This practice survived in the form of periodic lectures for new workers. By the end of the war these lectures had provided the basis for a handbook on radiation hazards and preferred techniques for coping with them.[39]

In August 1944 Los Alamos plans to forgo biological research were scrapped. Waiting for research results proved just as frustrating as waiting for instruments. Chicago and Rochester had joined Berkeley since spring in trying to define the acute and chronic toxic effects of plutonium. Despite much hard work, useful results came slowly. Biological research could not be hurried and tight security hampered the exchange of what data there were. Hempelmann and others concerned with radiation safety at Site Y grew restive. For whatever reasons, the project failed to meet their pressing demands for hard data on which to base safeguards. An accident in August 1944 transformed dissatisfaction into action.[40]

Ten milligrams of plutonium exploded in a chemist's face. Although only a chemical mishap, it caused him to swallow an unknown amount of plutonium. Kennedy and his colleagues were shocked to learn that the Health Group could neither measure how much lodged in the chemist's body nor predict how severe the outcome might be.[41] Hempelmann, sharing the concern, promptly requested Oppenheimer to authorize research on certain "medical problems . . . to which answers are urgently needed." He proposed three key questions: How could plutonium in human urine and feces be measured? What was the ratio between excreted and retained plutonium? How could plutonium in the lungs be measured? The answers obviously required human subjects, but Hempelmann also wanted a substantial animal program to support the search.[42]

Oppenheimer endorsed finding techniques to detect and measure plutonium in humans. That seemed clearly useful, addressed as it was to the direct medical needs at Los Alamos. About the more purely biological research, however, he remained doubtful: "I feel that it is desirable if these can in any way be handled elsewhere not to undertake them here."[43] Oppenheimer reviewed the problem and Hempelmann's research plan for the Los Alamos Administrative Board on 17 August. He voiced distress "that other parts of the project are not attacking this problem inasmuch as it will be fatal to some people while others will be taken off their work when there is nothing wrong with them."[44] Informed of these views, General Groves sent Chief Medical Advisor Stafford Warren to Los Alamos. Oppenheimer, Kennedy, and Hempelmann conferred with Warren on 25 August.[45] They agreed on a "biological research program . . . with relatively high priority" along the lines proposed at Site Y.[46] It began at once.[47]

Plutonium decays slowly—its half-life is 24,400 years—and emits only alpha particles and very weak X rays. These qualities prevented directly detecting it in the body and dictated the main task, finding it in urine. A key problem was the minute amount—a trillionth of a gram or less per liter of urine. Wright H. Langham, who had joined the Met Lab in 1943 after winning his Ph.D. in biochemistry from the University of Colorado, came to Los Alamos and headed the research effort. By January 1945 the basic problem had been solved. Initially, however, the new method saw only limited use. Collecting proper samples required special facilities and rigorous techniques to exclude any chance of contamination. The animal studies that provided the basis for the new test also needed cross-checking with tracer studies on human subjects. These began at both Los Alamos and Chicago in April, a little later at Berkeley. All this took time. Testing of workers began in spring 1945 but became routine only that fall. Even then, it required too much time and effort to be done often. Nasal swabs remained useful for spotting the most heavily exposed workers, who then had more frequent urinalyses.[48]

Nose swipes also remained the only basis even for guessing at lung burdens of plutonium. Although finding a precise test for that purpose stood high on the Health Group agenda, the effort failed. One reason was limited resources. Researchers, however, also lacked any clear-cut notion of how to proceed. In contrast with the technically demanding but straightforward task of measuring plutonium in urine, no useful approach to the lung problem suggested itself. It remained unsolved. So did the problem of plutonium detoxification. Once lodged in the body, how could plutonium be removed? Initially of little concern at Los Alamos, this question arose only after the Health Group had devised the urine test and began checking workers early in 1945. Then the group learned that some workers might already be carrying too much plutonium in their bodies. Chemical methods of flushing plutonium from the bodies of exposed workers were under study elsewhere in the Manhattan Project. From the Los Alamos viewpoint, that was not nearly enough; once again, however, urgent pleas brought little help.[49]

The Health Group also played a role in guarding legal interests of both workers and project. That required keeping complete records: hazard studies, exposure logs, accident reports, and test results. The task should also have included routine tests and examinations of workers on the project and complete exit examinations for those leaving. But record keeping and testing demanded time and people stretched too thin. Hempelmann and his group closely watched work where hazards were clearly prolonged or severe. They could seldom, however, maintain complete records of nonexposure for those employed outside the danger areas. They also were hard pressed to cover those, such as cyclotron users, whose work involved constant but modest hazards. Hempelmann knew that "we do not have important legal evidence in case of future claims against this project." Partial blood records and unreliable urine assays for exposed persons also might have legal repercussions. Little could be done. Better records might be worthwhile, but radiological safety itself came first.[50]

LOS ALAMOS REDIRECTED

Until summer 1944 fission-bomb design centered on a gunlike device. A slug of fissile metal below critical mass—that is, too small to sustain a chain reaction—would be fired into a second subcritical mass hollowed to receive it. The combined mass would then be great enough to explode. This procedure remained the basis for the uranium bomb. The method was simple and certain, an advantage that outweighed its relatively extravagant use of fissile metal. Development of the uranium gun-type bomb proceeded apace. For physical reasons—confirmed only after plutonium became available for research during the first half of 1944—the gun could not be used for a plutonium bomb. Uranium- and plutonium-bomb designs parted company. The final decision to scrap the plutonium gun came on 17 July 1944. Los Alamos turned to implosion, a much less certain technique.

The theory was simple. Conventional explosives would enclose a sphere of plutonium. Detonated inward simultaneously and symetrically, they would compress the sphere to a critical mass. This was implosion: the force of chemical explosives focused inward to compact the plutonium, which would then sustain fission and explode. Discussed and studied since the early days of the project, the method held real promise, notably of efficiency. If it worked, less metal required per bomb could mean more bombs sooner. Making it work was something else again. Chemical explosions presented still another poorly known area of physics which demanded more research. The risk of failure had seemed serious enough to place implosion well behind the gun as basis for a bomb. Despite months of work, most of the hard questions remained unsolved. But other options were gone. Implosion seemed the only way to salvage the huge investment already committed to plutonium. Los Alamos underwent a complete reorganization as it shifted to the new implosion technique.[51]

Reorganization of the Chemistry-Metallurgy Division most directly affected the Health Group, chiefly because the chemists faced the biggest risks at Site Y. In September 1944 the monitoring and decontamination group (CM-1) acquired a new leader, Robert H. Dunlap, and transferred some of its members to the Health Group. CM-1 now conducted only routine monitoring and decontamination. The Health Group would set standards and approve procedures. Hempelmann expected to receive daily written reports and to provide help for any unusual problems. Responsibility so divided, however, soon had the two men at odds, and the smooth working relationship of the preceding six months gave way to growing friction. Matters reached a crisis in January 1945, marked by lengthy and acrimonious memoranda from both Dunlap and Hempelmann to Kennedy. Before the dispute could be resolved, a worse crisis intervened.[52]

In January 1945 a serious fire broke out in one of the shops. It raised the specter of a fire in Building D which might strew plutonium over much or all of Site Y. Never very safe to begin with, the building also approached its limits in handling still growing amounts of plutonium. Planning for new facilities less vulnerable to fire and better designed to house plutonium and polonium safely began at once. Alpha contamination problems meanwhile went to a new health instruments group (CM-12) under William H. Hinch. Acting quickly to reduce friction, he soon had working relationships with the Health Group back to normal. The alpha problem persisted, however, and months passed before it could be well controlled. It was resolved only when the chemists moved into their new laboratories in the fall. Well before then, however, other and potentially far greater hazards claimed an ever growing share of Health Group attention.[53]

The concerted attack on implosion made work at Los Alamos still more hectic, further taxing the Health Group's already strained resources. Although growing to meet new demands since early 1944, it never caught up. Chronic understaffing, in Hempelmann's view, marked the Health Group throughout the war.[54] Some of the burden could be transferred. Divisional groups patterned on CM-1 took over most of the routine monitoring. The Health Group largely ceased trying "to advise specialists in their own fields as to safe operating procedures." Instead, it confined itself to making certain that the experts grasped the dangers and to checking the safety rules they adopted. The Health Group's often-expressed reluctance to conduct its own research reflected these circumstances. Research seemed something of a luxury when the group could not even keep fully informed of, much less regularly inspect, all technical operations at Site Y.[55] Implosion studies multiplied the problems.

One means of coping with hazards was keeping them at a distance. Potentially risky programs left the mesa for nearby canyons. Relocation limited danger but also made it more difficult for the Health Group to monitor such programs. The first chain-reacting unit built at Los Alamos was the so-called water boiler, a reactor powered by enriched uranium dissolved in water. Located at omega site in Los Alamos Canyon, it first ran in May 1944. Heavy concrete shielding, remote controls, and low power made it only a minor hazard. Redesigned as a strong neutron source, however, it ran again in December. To make decontamination less often required, the new design included an air-flushing system for gaseous fission products. It worked well, but not perfectly.[56] Leaking exhaust gas lines twice caused overexposures deemed "negligible, less than .5r," to seven chemists. When the water boiler did require decontamination, five chemists were exposed to the highly active fission products. According to standard Victoreen pocket chambers, "personnel working on the renovation received from 1 to 5 daily doses [0.1–0.5 roentgen] per day. The average dose, figured for the entire two weeks of renovation probably consisted in about 2 daily doses [0.2 roentgen] per day per person." Workers did not, concluded the group leader, "receive a dangerous dose."[57]

Bayo Canyon housed the far more hazardous RaLa program. RaLa was radiolanthanum, a relatively short-lived fission product and a strong source of gamma rays. The intense rays allowed instruments to chart and measure the course of test implosions. A group from the Weapon Physics Division conducted the experiments. Doubts about how much contamination to expect from an imploded source equal to hundreds of grams of radium dictated the remote site. Uncertainty also prompted the physicists to work closely with the Health Group in planning and running the first trials during September and October 1944. Members of the group observed the tests firsthand from sealed army tanks. Permanent structures rose only after the hazards proved conveniently small.[58] The greater risks, in fact, fell to a second group, the chemists who extracted radiolanthanum from the mixture of fission products shipped to New Mexico from Tennessee. Single batches might total as much as 2,300 curies. Working bugs out of the remote-handling system took six months.[59] During that time failures and accidents periodically exposed chemists to "considerably more radiation dosage than was desirable."[60]

The critical assemblies group shared omega site with the water boiler. Headed by Otto R. Frisch, its purpose was to find just how much enriched uranium the first bomb would need. The expatriate Austrian physicist had played a major role in early British work on a fission bomb and joined the British contingent assigned to the American development program.[61] No one doubted that the nearly ready uranium gun would work, but no one had yet seen uranium explode, either. When the first large amounts of enriched uranium reached Site Y, Frisch proposed an experiment. Blocks of the metal hydride would be assembled as a bomb. A large hole left in the center would preclude a chain reaction. The missing core, mounted on rails, could then be dropped through the hole.[62] For a split second it would just barely create the conditions for an explosion, as near an approach as could be conceived, Frisch remarked, "towards starting an atomic explosion without actually being blown up."[63]

When Frisch formally presented the plan, a chuckling member of the council likened it to "tickling the tail of a sleeping dragon."[64] And so it became the dragon experiment. Set up in a matter of weeks, it was completed on 18 January 1945. The pace was hectic because the material had to be returned to the weapon makers. Enriched uranium was in very short supply, and the hydride had to be reduced to metal for use in the real bomb. Despite haste and pressure, the test was perfect.

"Everything happened exactly as it should," Frisch recalled. "When the core was dropped through the hole we got a large burst of neutrons and a temperature rise of several degrees in that very short split second during which the chain reaction proceeded as a sort of stifled explosion."[65]

Other criticality tests followed, although later trials rarely shared the original dragon's flamboyance. Paradoxically, however, in some ways they presented even more serious dangers. The reasons were largely psychological. Proper care precluded any danger at all; nothing could happen unless an assembly exceeded the critical amount. A long series of trouble-free tests could foster a degree of overconfidence. "Those of us who were old hands felt impervious to the invisible danger," a member of the critical assemblies group recalled. "I am afraid that familiarity indeed breeds contempt of danger."[66] No one, of course, ever intended to be careless, but the special risks of any particular test could not always be predicted. Criticality might follow even a minor mistake, and when it happened, it was very sudden.[67]

The first serious mishap with critical assemblies occurred during a safety test. Experimenters slowly added water to enriched uranium to study the hazards of accidental flooding. Criticality came sooner than expected. Before the water drained, the reaction became intense. Four workers suffered acute exposure to large amounts of gamma and neutron radiation. "No ill effects were felt by the men involved," according to the official history, "although one lost a little of the hair on his head."[68] Two postwar accidents, both to veterans of the program, had less happy outcomes. In the first, Harry K. Daghlian was working alone on the evening of 21 August 1945. Onto an almost complete assembly he inadvertently dropped a block of the tamper material used to reflect neutrons. Criticality resulted at once. Daghlian died of acute radiation syndrome less than a month later. Louis Slotin, the second victim, had won his doctorate in physics from the University of Chicago after serving with the Abraham Lincoln Brigade in the Spanish Civil War. At Los Alamos he became leader of the critical assemblies group. In May 1946 the screwdriver he was using to prop one piece of fissile metal away from another slipped. A burst of radiation washed over all eight men in the room before Slotin was able to knock the piece away. Slotin died nine days later, but his quick action saved the other seven men. Farther away and partly shielded by Slotin's body, they were all overexposed, but they all survived.[69]

EARLY TRINITY PLANNING

The largest experiment was actually test-firing an implosion bomb. It was also the experiment most clearly fraught with danger, not only to Los Alamos workers themselves but also to the public. Planning began in March 1944, although such a test had not then seemed urgent, for the gun still seemed to be the best method for the plutonium as well as the uranium bomb. The Ordnance Engineering Division formed a group that, among other duties, began to prepare for field-testing an atomic bomb. Kenneth T. Bainbridge became group leader. A physicist, Bainbridge had left his teaching post at Harvard to work on radar development at the MIT Radiation Laboratory before coming to Los Alamos in 1943. When implosion became the focus of the Los Alamos effort in mid-1944, Bainbridge's group shifted to the Explosives Division. Too much remained unknown to permit real test planning. Indeed, demands for basic studies of detonation sharply reduced the test-planning effort through the last half of 1944 and early 1945.[70]

Much groundwork had nonetheless been laid. One section of Bainbridge's group focused on instrumentation: what to measure, how best to measure it, what special instruments to use. A second section concerned itself with a fizzled nuclear explosion. If imploded plutonium failed to explode, the costly and hazardous metal might scatter far and wide. The answer was a steel cylinder 25 feet long and 12 across. Fifteen-inch thick walls would, if the figures were correct, contain the nonnuclear explosion. Nicknamed Jumbo, the 214-ton vessel was designed, built, and installed at the test site, but it was never used. The more one learned about implosion, the less urgent seemed something like Jumbo. Perhaps more to the point, using it would hamper the very measurements for which the test was chiefly designed.[71]

A second scheme, nearly as bizarre, proved more fruitful. The best measure of the bomb's efficiency, planners believed, would be the ratio of unchanged plutonium to fission products in the postblast debris. For best results, samples of the ground must be collected as near the point of detonation and as soon after the event as possible. Some means of reaching the highly radioactive bomb crater promptly yet without undue risk to the samplers had to be found. Enter the two Sherman tanks already at Site Y for the RaLa program. With proper fittings, a tank might both cross the rough terrain and protect the sampling crew from overexposure. It was a big job. One of the tanks received a lead wrapping, 5.5 tons in walls 2 inches thick around the crew compartment. Fully airtight, it also carried a self-contained air supply. External sampling devices beneath the tank but controlled from within completed the fittings.[72]

Perhaps the most important early decision was choosing the test site. The search started with Bainbridge poring over maps during spring 1944. Producing useful data from the test depended on proper terrain and climate, but they were not the only criteria. The site had to be near enough Los Alamos for workers to travel back and forth without undue trouble. Transporting material and supplies also had to be reasonably easy. Yet the site had to be far enough from Los Alamos to avoid easy linkage to the secret laboratory. Security also dictated a remote and sparsely peopled region. No one could be allowed to remain on the site except those engaged in the test; the fewer that would have to be removed, the better. Safety vied with security as a motive for seeking a site with few people nearby. "The area had to be remote," Bainbridge recalled, "so that people could be evacuated." Even a fizzled bomb still might scatter hazardous amounts of plutonium some distance. If the test succeeded, the problem could become "dangerous radioactive fall-out." Either outcome might require action. Once again, the fewer to be moved the better.[73]

Eight sites showed promise. On-the-spot surveys by car, jeep, and low-flying airplane narrowed the choice to two. One was the Jornado del Muerto region, a desert valley on the Alamogordo Bombing Range (today the White Sands Missile Range) in southern New Mexico. The other was an army tank-training ground in the Mojave Desert of southern California. The nod went to Alamogordo in September, specifically to an area 18 by 24 miles in the range's northwest quarter. It was 12 miles from the nearest ranch house and 27 from the nearest town.[74] The time had come for a name. Obscurely inspired by the seventeenth-century devotional sonnets of John Donne (and perhaps by Hindu mythology as well), Oppenheimer suggested Trinity. That became the name both of the site and of the test conducted there.[75]

Work on Trinity base camp began at once. At the end of 1944, major building completed, a military police unit moved in as guards and custodians. Like Site Y, Trinity far outgrew its planned limits. The intended total of 160 military and civilian personnel tripled. By mid-July 1945 the site housed 250 scientists and technicians, plus an equal number of soldiers. Spectators swelled the total to 700 the weekend of the test. But that was months away. As 1945 dawned, the question of a test remained open because key research on bomb physics remained unfinished. Things looked better by the end of February. Although the gap between laboratory findings and a working bomb remained wide, an acceptable design could be set in principle. A full-scale test of the device now seemed feasible. It also appeared mandatory. In contrast with the gun-type bomb, planners judged the chances of failure too large to risk the first trial of an implosion bomb in combat. Oppenheimer formed Project TR (for Trinity) to conduct the test. He assigned it divisional status and kept Bainbridge in charge.[76]

The Health Group had just reorganized. Although still understaffed in Hempelmann's view, it had grown large enough to warrant a more formal structure. James Nolan left the post hospital to become alternate Health Group leader. Three sections were formed. Wright Langham directed biomedical research. The Explosives Division provided Joseph G. Hoffman to head the health physics section. Holder of a Cornell Ph.D. in physics, Hoffman went to Los Alamos in 1944 from the National Bureau of Standards. Army Medical Corps Lieutenant J. H. Allen took charge of the Health Group's new clinical section. A growing share of Health Group effort in the months that followed centered on safety planning for Trinity.[77]

Radiological safety played only a modest part in early Trinity planning. When no one knew how to make a bomb or if it would work, safely testing it hardly loomed as a central concern. Yet that was not the whole story. In the early years almost everyone perceived fission chiefly as the source of a uniquely powerful explosion. Safety then seemed mostly a matter of protection against blast and bomb fragments. These dangers, of course, were nothing new, however much larger the scale. That a fission bomb might also pose unique hazards attracted little notice, although some warnings were voiced.[78] One of the first came from the British M.A.U.D. Committee in July 1941. Its final report concluded that a uranium bomb could be developed before the war ended, a key factor in the American decision to launch the Manhattan Project. Still, a fission bomb would produce, the report also noted, "very large amounts of radioactive materials." Scattered widely by the blast, they could pose severe long-term hazards. "The physiological effects of these radiations are delayed and cumulative," it warned, "so . . . great care would have to be taken in working anywhere near."[79] A Los Alamos committee also noticed the "danger of spread of radioactive fission products" from a bomb.[80] Radiological safety began to emerge as a question, however, only when Oppenheimer committed Los Alamos to the test.

SAFETY PLANNING AND THE 100-TON SHOT

Project Trinity had no sooner been formed than a self-appointed committee began to discuss "the hazards related to the test." It included Hempelmann and electronics expert Richard Watts, as well as several physicists. Bainbridge and his chief aides joined the second meeting. They discussed three main topics: "(1) danger to personnel at the site and in the neighboring areas during and after the shot, (2) medicolegal aspects of these hazards and (3) instruments and organization needed to cope with the above hazards."[81] Most concerns about dangers from blast, fragments, heat, light, and initial radiation—that is, gamma rays and neutrons emitted during the explosion—were quickly settled. The nearest test workers would be amply sheltered behind earth, timber, and concrete 10,000 yards, almost 6 miles, from the zero point. No member of the public would be closer than 20 miles, well beyond any danger. Residual radioactive debris was the real hazard.[82]

In these March 1945 meetings, the deepest concern was focused on the results of an inefficient implosion—plutonium merely converted into fine dust. Almost 6 miles away at the closest test shelters, a person might inhale the 1-microgram tolerance dose in eight minutes. Since wind-borne dust tended to disperse, reaching that dose would take eight hours at the nearest town. It was unlikely, however, that the hazard would last that long, as the dust drifted on, "unless some form of precipitation might result in the deposition of the entire cloud on the town."[83]

Efficiency altered the nature of the hazard. Plutonium would fission into other radionuclides. Neutrons striking elements in the ground nearby would turn some into active nuclides. Activated debris would be swept with fission products into the expanding fireball and lifted high into the air. They would form a cloud of intense gamma and beta emitters drifting on the wind. The very efficiency of the blast, however, promised to reduce the danger. Thousands of feet above the ground, the cloud would present no direct threat to people below. It also would be far more diluted by air than a cloud of plutonium dust.[84]

Largely ignored in the March talks was the question of fallout. As turbulence subsided and the cloud cooled, fission products and active nuclides would condense and plate out on cloud-borne dust. Radioactive dust would then begin to drop in the cloud's wake. Where and how much would depend on the precise characteristics of the blast and on specific weather conditions. This phenomenon was then poorly known, but fallout seemed a minor hazard, as fission products swiftly decay. Perhaps more to the point, doubts about implosion channeled thinking toward the risks of failure; plutonium dust seemed a more likely outcome than a fallout cloud. That changed as confidence in the gadget grew. Meanwhile, though, coping with the hazards of plutonium contamination looked more urgent than dealing with fallout.[85]

Medicolegal questions about Trinity were easy to raise but hard to answer in March 1945. The Health Group fully intended to monitor the hazards, not only on the test site but also in nearby towns and across the region a cloud might traverse. The readings, Hempelmann hoped, would provide "permanent records . . . for future reference" against any legal action.[86] Even this limited goal might be hard to achieve. Readings required instruments, but most of them remained in short supply and some did not yet exist. Worst served was the hazard then deemed most likely: alpha-active dust from scattered but unchanged plutonium. "No instrument has been developed as yet which will give an instantaneous reading of the alpha activity of the air," Watts observed in the March meetings.[87] Instantaneous reading was crucial if monitors were to warn workers promptly of threats at the three test shelters. Portability was just as crucial for tracking the cloud cross-country.[88]

Watts and his health-instruments team now worked full time for the Health Group. Much of their effort focused on a portable alpha counter. The new device could detect as little as 100 counts per minute per liter of air, enough, Watts noted, "to warn personnel of levels of activity . . . dangerous to breathe over a period of 100 minutes."[89] The new counter was of the proportional type. Gas amplification was enhanced by using methane instead of air and by having the gas flow through the chamber to lower its pressure. The methane tank and heavy batteries made the counter bulky, but less need for electronic amplification outweighed that drawback. The result was a simple and rugged device, heavy but otherwise well suited for use in the field.[90]

Monitoring towns posed less critical problems. Portability mattered little and distance reduced the hazard. The main reason, however, was

Sneezy, the nickname for an alpha air sampler devised by the Chicago instrument group. It used a simple and sensitive technique. The sampler drew air at a metered rate of 50 liters per minute through a large piece of special fine-fibered filter paper. Monitors then measured filter activity in a fixed counter. Los Alamos usually called it the Filter Queen, after the commercial vacuum cleaner adapted to draw air through the filter.[91] Slowness was its one real drawback. Atmospheric radon had to be accounted for, usually by letting filter activity decay for a few hours before taking a second reading. The process became quicker when a rough correction factor mathematically adjusted the observed reading for excess natural activity. The method still imposed waiting—the time required to draw enough air through a filter and then transfer it to the counter. The delay seemed no very serious matter in a system regarded as mainly precautionary in any event. The Filter Queens installed at Trinity base camp and the nearest towns served the purpose.[92]

If Trinity succeeded, fission products rather than plutonium would pose the hazard. Monitoring would then have to center on gamma and beta activity rather than alpha. In contrast with alpha meters, beta-gamma meters did not have to be produced from scratch for the Manhattan Project. The only problem foreseen for Trinity concerned the wide range of activities. The answer called for two meters, both in commercial production. To cover the lower range, up to 10,000 counts per minute, Watts proposed the Hallicrafters Model 5 Geiger-Müller counter. For higher ranges, up to 10 roentgens per hour, he suggested the Victoreen Model 247 ionization chamber.[93]

Geiger-Müller counters were not so useful then as they later became. The tube's electron avalanche pulses are not proportional to the ions that trigger them; the counter therefore cannot provide a direct measure of rate or total ionization. Ionizaton merely produced pulses, resolved and integrated as meter readings. Such readings could be calibrated as exposure rate for a defined energy spectrum, as from a sample of fission products. This was, in fact, done. Ionization events at too rapid a rate, however, presented an unsolved problem; they caused continuous discharge in the tube and decreased meter readings, even at modest ionization levels. This difficulty tended to restrict their use in the field, where concern centered on levels high enough to pose a direct health hazard. The Health Group found Geiger-Müller counters of more value for record keeping. Connected to recording devices and installed at the base camp and nearby towns, they provided a permanent legal record of low-level gamma activity.[94]

The Victoreen meter emerged as "the most useful instrument for field work."[95] Basically an ionization chamber, it required no electronic amplification. It was a touchy survey meter, but it was also light and compact. Its greatest advantage over Geiger-Müller counters for fieldwork, however, lay elsewhere. Measuring total current rather than pulses, it handled high exposure rates better.[96] In practice, the supply of survey meters proved more limited than Watts had expected. As neither Chicago nor the manufacturers could meet the demand, Watts's group built a number for Trinity. Even then, Watts had to reduce the projected sixty instruments in the field to thirty. The lower number "was barely found to be sufficient" to equip the test shelters and the mobile units.[97]

Personnel dosimetry also figured in Health Group planning, but it presented no problems. Each person present for the test would wear a film badge. By spring 1945 film badges had become standard and readily available in any numbers that might be needed. Most workers would also carry pocket dosimeters; these, too, had become standard and plentiful.[98] Hempelmann and Watts outlined their Trinity plans for Kenneth Bainbridge and Stafford Warren on 12 April 1945. Bainbridge as Project TR director and Warren as chief medical officer of the Manhattan Engineer District approved the plans. Approval was largely a formality. Bainbridge and Warren, in effect, merely validated an effort already well under way.[99]

Health Group plans faced their first trial in the test rehearsal scheduled for early May. No one knew much about the characteristrics of really big explosions or how to measure them. Conducting a large-scale test like Trinity without some kind of practice seemed chancy as well. For these reasons, Bainbridge decided on a trial run using 100 tons of high explosives. The plan was to rehearse the later test on a smaller, controlled scale. Workers stacked boxes of high explosives 16 feet high on a 20-foot wooden tower. Calculations suggested that this structure was a precise, scaled-down version of the gadget on its 100-foot steel tower. Instrument placement was likewise scaled to the expected yield of the nuclear blast, roughly equal to 4,000 or 5,000 tons of high explosives. Plastic tubes running through the stack held dissolved fission products to simulate the radioactivity of the full-scale test. Radiological safety inspired three major concerns: the process of spiking the stack with fission products, radioactivity on the ground after the test, and the fate of the cloud.[100]

The fission products came from a uranium slug irradiated at Hanford for 100 days. Trucked to Trinity late in April, the slug went into a stainless steel vat. The vat itself rested in a buried concrete shell lined and capped with lead bricks. Concentrated nitric acid dissolved the slug in a daylong process. The Health Group closely monitored worker doses. Heavy shielding kept radiation exposure below 0.1 roentgen per hour. As a further precaution, a stream of nitrogen swept the active gases to vent 1,000 feet from the work area, where a blower diluted them to safe levels.[101] "Exposure meters were worn and it can be definitely stated that no one received a dose in excess of 0.5r of radiation."[102]

Once dissolved, the fission products were pumped to the stack by remote control. There the solution lacked full shielding. Its 1,000 curies of beta and 400 of gamma activity posed a more severe hazard. Greater caution, however, largely avoided any problems until the day before the test.[103] Then, during final preparations, two workers suffered mild overexposures. One received about one-and-a-half times the daily tolerance dose. The second may have "received about three or four daily doses." Installing the detonator "required him to stay in the vicinity for about four hours. This dosage was not measured but was not considered serious because he has had no other exposure." After the test, workers covered the vat with fresh earth and surrounded the area with a guard fence.[104]

Hempelmann judged "the hazards relative to the 100 ton shot containing fission products at Trinity on 7 May 1945," to be "slight."[105] The shot went off just before dawn. Twelve miles away at the camp Stafford Warren saw the "red-brown to orange" flash, but "the light left only a momentary blind spot." Watching the "rising cloud of dark smoke," he heard "a sharp thumping explosion" echoed weakly from the surrounding mountains.[106] A small Health Group party drove to the scene. Masked, gloved, and booted, they conducted a quick survey and flagged the active area, a 60-foot circle centered on the tower. Blast devastated a much larger circle, 0.3 mile across. At 6:30, two hours after the explosion, Hempelmann judged the crater safe for several hours' exposure and opened the area to scientists eager to retrieve their data.[107]

Meanwhile a second party, also protectively garbed, rode the lead-lined tank designed to obtain quick samples near the point of blast. The trial helped resolve some questions: Could a tank traverse the debris of a bomb crater? How well did the shielding work? The test also offered a chance to try the new rocket sampler devised for Trinity by a team from the California Institute of Technology. The rocket was launched from a skid towed by the unshielded second tank. It carried a sampling tube and a scoop for 200 yards. Sampling complete, an attached steel cable reeled it back. Collected samples showed that 2 percent of the activity remained within 300 feet of the tower.[108]

The other 98 percent rose in the hot column of air created by the blast. Smoke, dust, and debris mushroomed into a cloud at 15,000 feet. As predicted, the day dawned bright and clear; unlimited ceiling and visibility made the cloud easy to observe. Proper weather for the test was a key planning factor. Meteorologist Jack M. Hubbard joined Trinity in April to study local weather patterns and provide the detailed forecasts that, in part, governed the shot's timing. He predicted perfect weather for the early morning of 7 May and his forecast hit the mark.[109] Observers at Trinity base camp watched the cloud ride a 30-mile-an-hour wind almost due eastward across the Sierra Oscura, passing between the nearest towns, Tularosa and Carrizozo. Four hours later it could still be seen drifting south of Roswell, New Mexico. No one, however, chased the cloud. Trinity planners saw "very little likelihood of any contamination ever reaching the earth . . . , [given] a dilution of 10,000 times for every 2,000 feet vertical descent of such clouds."[110] The blast itself escaped notice in nearby towns, although a forewarned observer 60 miles away detected its light and sound. Ground shock, however, was imperceptible even at the 10,000-yard distance of the test shelters.[111]

The test was not flawless. The remote site hampered efforts to procure and transport test gear, causing delays. Electrical interference between experiments produced problems, not all of which could be resolved. Never regarded as anything but a trial run, however, the 100-ton shot was judged a success. It clearly highlighted the need for improved test procedures; even more, it showed where the work must be done. Results amply repaid the effort, particularly data on amounts and distribution of fission products likely to remain on the ground.[112] This knowledge was vital for all Trinity goals, not least "for planning the . . . protection of personnel for the final test."[113] Some safety questions, however, the 100-ton shot was simply not designed to answer, especially the effects of less than perfect weather on radiological hazards. Such questions had legal and security, no less than health and safety, aspects. They began to emerge only when Oppenheimer decided to proceed with Trinity. In the weeks after the 100-ton shot, these questions came under ever closer study.[114]

4

TRINITY

SCIENCE, SECURITY, SAFETY

Secrecy cloaked the Manhattan Project from the outset. In the hierarchy of project goals it stood second only to making bombs that worked. Scientists frustrated by compartmentalization sometimes wondered if it ranked merely second. The army assigned each of the project's far-flung activities its own management box. Communicating between boxes ranged from difficult to impossible. At least some of the Health Group's problems with instruments and health data—to cite one small example—stemmed from such deliberately clogged lines of communication. Testing a fission bomb on American soil, however remote the site, clearly threatened the secret. The most violent man-made explosion in history could scarcely pass unseen. Elaborate public safety measures seemed likely only to render it more noticeable. Fortunately, safety and security did not have to exclude each other. To some degree, at least, the same measures keeping Trinity safe from prying eyes could help keep the public safe from the test and testers safe from lawsuits.[1]

Safety never commanded topmost concern at Los Alamos. Getting the job done came first. In testing the bomb, however, safety may have ranked even lower than normal. As James Nolan, chief safety planner for Trinity, later explained,

> Possible hazards were not too important in those days. There was a war going on. . . . [Army] engineers were interested in having a usable bomb and protecting security. The physicists were anxious to know whether the bomb worked or not and whether their efforts had been successful. . . . The bomb was designed as a weapon of warfare primarily utilizing blast and heat for destructive forces. . . . [R]adiation hazards were entirely secondary.[2]

Hymer Friedell, second-in-command of the Manhattan District medical office, commented more tersely: "The idea was to explode the damned

thing. . . . We weren't terribly concerned with the radiation."[3] Radiological safety, in other words, competed with other test goals; it ranked
higher than most, perhaps, but not highest.

Writing a quarter century after the event, the head of the Manhattan
Project stressed safety concerns in Trinity planning. General Leslie
Groves also showed how the overriding goal of secrecy linked to and
shaped those concerns. He listed six "immediate military requirements
[for] adequate security": (1) barring strangers from the test site; (2)
preventing harm to project members; (3) reducing chances that out-
siders could learn of the explosion; (4) safeguarding the public from
fallout; (5) planning for emergency evacuation; and (6) forestalling any
national press reports that might alert Japan. Not all these requirements
carried equal weight or demanded equal effort. The concerns did not
all arise at the same time and the list was, in any event, partial. Inevitably, Groves's "Recollections" reflected 1970 as well as 1945 views.[4]

His main point, however, was entirely valid. In 1945 single motives
seldom dictated actions, as running the test site itself exemplified. The
task called for juggling science, security, and safety, more often than
not in ways that supported one another. Lieutenant Howard Bush's
military police managed the site and provided support services. The
125-man guard force also protected Trinity from intruders, helping both
to forestall spying and to prevent hapless strangers from wandering
into danger. By keeping strangers out, the army also reduced its
chances of being sued for damages.[5]

When Groves visited Site Y in April 1945 for a briefing on Trinity
plans, his first questions concerned legal matters. Civilian structures
might suffer damage from earth shock and air blast. Had Los Alamos
taken proper steps to measure such distant effects? Validated records
could help secure the army against damage claims.[6] Groves told Kenneth Bainbridge, Trinity director, "that he was going to get additional
legal talent to consider and act on the legal aspects of TR tests."[7] Groves
also asked Los Alamos Director Robert Oppenheimer for a "written
opinion as to the safety of the proposed test from the points of view
of earth shock, air blast and toxic effects."[8] Oppenheimer professed no
worries about shock and blast. Neither, he assured Groves, posed any
"danger whatever of damage . . . to property or personnel outside the
area under our control."[9] Groves, however, remained cautious. On the
day of the test, army intelligence bolstered Los Alamos efforts with
twenty agents stationed in towns up to 100 miles away. Equipped with

recording barographs, they carried orders to secure legally valid data on remote shock and blast.[10]

A prospect of really large-scale catastrophe lingered in the background. Secrecy breeds rumor and speculation. Whatever scientists might say, the first test of a radically new bomb provoked unease.[11] At Trinity base camp on the eve of the test, "I was furious to hear," Bainbridge recalled, "discussions of the possibility that the atmosphere might be detonated. This possibility had been discussed at Los Alamos and had been quashed by intensive studies . . . by Hans Bethe and others."[12] The idea was more than two decades old. In 1920 Arthur Eddington had suggested that the sun's energy might derive from hydrogen fusion. Two years later, Francis William Aston noted the likely value of controlled fusion on earth; he also sounded a warning. Once started, fusion might escape control. "In this event, the whole of the hydrogen on the earth might be transformed at once and the success of the experiment published at large to the universe as a new star."[13]

Apprehension ran deep in the ranks. A young army technician assigned to the Special Engineering Detachment at Trinity knew none of the bomb's secrets. But he recalled his prewar reading of

> the little book by Ernest Pollard and William Davidson on *Applied Nuclear Physics*. One could not read the chapter on nuclear fission without being struck by a melodramatic remark uncharacteristic of physicists: "If the reader wakes some morning to read in his newspaper that half the United States was blown into the sea overnight he can rest assured that someone, somewhere, succeeded."[14]

In work that later won a Nobel Prize, Hans Bethe defined the process of stellar fusion just before the war. At Los Alamos he and others, notably Edward Teller, studied methods for using fission to trigger a fusion bomb. They quickly learned just how difficult fusion would be.[15] No one could flatly reject the chance of runaway atomic energy, but as Bainbridge wryly observed, "We all put our faith in Bethe."[16]

Catastrophe on such a scale, in any event, surpassed the limits of human planning. Lesser disasters, however, did not. Groves met New Mexico's governor in secret to arrange emergency measures; he received authorization to declare martial law "over as large an area as might be necessary."[17] *New York Times* reporter William L. Laurence had just joined the project. Atomic bombing of Japanese cities promised a big story, and Groves wanted at least one writer to know the background. Meanwhile, however, Laurence's skills had other uses.

Groves asked him to write a series of press releases on successively more disastrous outcomes and wider areas of martial law. The proper code word transmitted from the scene would tell Groves's secretary in Washington which statement to release.[18] With these measures, Groves recalled, "I had provided for what I considered to be impossible: that the destructive force of the bomb would be many times our maximum estimates."[19]

Fallout remained the most realistic danger. Joining theory to data from the 100-ton shot of 7 May gave meteorologists, physicists, and doctors a good idea of how the cloud would behave and of the best weather for the test. They noted three key factors. First came strong westerly winds aloft to blow the cloud in its most active phase past the closest dwellings. Moderate turbulence would help disperse and dilute the cloud. Finally, the test should be postponed if rain- or thunderstorms seemed likely to wash fission products from the air over too small a region. Conditions for the final test, in brief, should match the nearly perfect weather of the 100-ton shot.[20]

Safeguarding the public against fallout demanded good weather forecasts. Groves transferred Phil E. Church from the Metallurgical Project to bolster Jack Hubbard's Trinity team. Church had worked at both Chicago and Hanford on problems of fission products released to the environment. Groves also enlisted advice from army weather experts for the test. Instrumented aircraft, weather balloons, radar, even smoke pots on nearby heights—all helped define local weather patterns. Military weather stations provided worldwide data for context.[21] With the right weather, "even the most extreme assumptions indicate that no community will be exposed to lethal or serious doses of radiation," Oppenheimer assured Groves; "it is my opinion that no personnel outside of the area controlled by us will in fact be measurably exposed."[22] As late as 2 July, Bainbridge insisted that Trinity demanded perfect weather: "If those conditions are not met exactly, the operations will be postponed until suitable meteorological conditions exist."[23]

HEALTH GROUP PLANS

Louis Hempelmann and the Health Group likewise assumed that the final test would wait upon proper weather. Just after the 100-ton shot, James Nolan, as Hempelmann's deputy, took charge of safety planning for Trinity. Plutonium-handling problems then peaking at Site Y kept

Hempelmann from a more active role in this stage of Trinity planning. Nolan spent a growing portion of his time at the test site. So did Richard Watts, leader of the health-instruments team attached to the Health Group.[24] By 20 June, Nolan had completed a plan of operations for coping with the "Medical Hazards of TR #2." Emergencies aside, the Health Group expected to advise, not command. Nolan's plan called for the Health Group

> to anticipate possible dangers to the health of scientific personnel, residents of nearby towns, and of casuals; to provide means of detection of these dangers; and to notify proper authorities when such dangers exist. It is also necessary to obtain records which may have medico-legal bearing for future reference. . . . [N]o person should (of his own will) receive more than five (5) r. at one exposure.[25]

Recovering data took priority over normal safety standards. The familiar limit of 0.1 roentgen per day still applied for routine work, but a special, higher limit guided urgent postshot activities. This practice became the norm for later test programs as well, although rarely in so discretionary a form as at Trinity.

Paul C. Aebersold, a new Health Group member, directed on-site radiation safety. Berkeley-trained, he had remained at the Berkeley Radiation Laboratory after his 1938 physics Ph.D. until coming to Los Alamos. Army doctors and technicians bolstered Aebersold's Trinity team. Posted at the base camp and at each of the forward test shelters— the bunkers named S 10,000, W 10,000, and N 10,000 for their direction and distance in yards from zero—they would monitor test workers, record exposures, and survey the site for contamination. Trinity workers clearly faced the test's most serious risks, but just as clearly they enjoyed the fullest protection.[26] Knowledge provided as sure a shield as special clothing or concrete walls. If by some mischance dusted with fallout, "they would be better prepared to take intelligent action . . . than would uninformed citizens."[27] To Groves, looking back on the hectic days before the shot, the key safeguard was "the issuance of the necessary instructions to all concerned. . . . [W]e were quite sure our instructions would not be blandly ignored, for everyone knew we were dealing with the unknown."[28]

Much the worst on-site hazard attended the postshot "entrance of personnel into the danger area" near zero to retrieve data.[29] Aebersold ringed the area 2,000 to 3,000 yards from zero with red warning flags on 6-foot poles 50 feet apart; appreciable ground contamination should extend no farther. Ionization-chamber robot sentinels lined the main access roads, spotted at distances between 400 and 10,000 yards from zero. During the test they provided remote readings throughout that region while workers remained under cover. Immediately after the shot, monitors wearing coveralls, booties, gloves, and gas masks would survey the ground for radioactivity and mark safe limits. Using these data, a "Going-in Board"—Bainbridge, Hempelmann, and Victor F. Weisskopf, consultant on nuclear physics measurements and radiation problems—would decide who could enter the danger area, and when.[30]

As the sample-collecting tanks had to enter promptly, their crews presented special safety problems. Herbert L. Anderson took charge of this mission when he arrived from Chicago early in 1945. He drafted an operational plan designed to move the tanks in and out as quickly as possible. Monitors would see Anderson's men to their tanks and check their gear. Once aboard, however, they were on their own.[31] Other entry parties also shared responsibility for their own safety. Scientists anxious to retrieve their data quickly appeared willing to pay the price of high exposures. The Going-in Board, however, intended to restrict entry until the radiation picture became clear. Health Group teams would provide clothing, film badges, and survey meters at S 10,000, the main entry point. They would show entrants how to use the gear and check them before leaving.[32] The Health Group, however, could merely urge cooperation. A "maximum dosage [which] should not exceed 5 R units" was strongly suggested, not ordered, though each entrant signed a statement acknowledging that "this is the individual's responsibility."[33]

Radiological hazards beyond Trinity's borders caused little concern until just weeks before the test. Fallout simply appeared to be a minor problem. Responsibility for off-site monitoring fell to Joseph Hoffman, Hempelmann's chief health physicist. Los Alamos "plans to send out radio equipped cars provided with instruments for measuring alpha particle and gamma ray intensities in outlying regions" met Groves's approval. "On the basis of these measurements, evacuation of inhabitants could be carried out if necessary."[34] Evacuation planning, however—to say nothing of putting such plans into effect—clearly threatened project security. Groves dismissed any thought of advance warning to nearby ranchers and townsfolk. Much talk but little action marked the subject through the spring. The danger seemed modest given the proper weather.[35] Groves proposed merely to station someone at the scene "to supervise the aspects where military action [evacuation] is desirable or might be necessary."[36] When Nolan handed Groves the Health Group plan on 23 June, he thought Groves more concerned about security than about safety.[37]

Just before Nolan finished drafting the Trinity safety plan the picture altered sharply. Joseph O. Hirschfelder and John L. Magee issued revised fallout figures. Physicists in the Theoretical Division, they studied postfission effects, that is, what happened to the products of a nuclear explosion after fission ceased. Probing such little-known phenomena, they found more questions than answers: How did the fireball form and rise in the air? How did matter within the fireball condense? What happened to debris swept into the fireball? How did particles disperse and fall? Their first short fallout paper relied chiefly on data from the 100-ton shot. To these they added some frankly pessimistic assumptions about the weather. The results were sobering. Hirschfelder and Magee concluded that fallout hazards might become more severe and extend over a much larger area than anyone had thought.[38]

Skepticism greeted these findings. Perhaps Hirschfelder and Magee had gone too far. Some critics noted that "only a most unfortunate combination of circumstances could lead to such a deposition of active material that it might prove disastrous."[39] Others wondered whether the figures showed an undue hazard. Computing results for the worst case, Hempelmann and Nolan found a total gamma dose of 68 roentgens over 14 days. They thought it "extremely improbable that injury would result in the case of people not previously exposed."[40] Questionable though the new estimate of hazards might seem—Hirschfelder and Magee, in fact, soon lowered it—evacuation plans began to look prudent, the more so when planners received another shock. Chances that Trinity could wait for perfect weather largely vanished.[41]

Hubbard's long-range weather forecasts suggested 18 or 19 July as the best time for the test. His second choice was 12–14 July. Early in June the Trinity schedules committee picked 13 July. By the end of June, however, technical problems imposed several days' delay. Based on weather, the firing date should then have been 18 or 19 July, but weather no longer controlled. The Potsdam Conference would open on 16 July. President Truman's meeting with Allied leaders to discuss Japanese surrender terms assumed the status of a deadline for Trinity. Los Alamos found itself under intense pressure from Washington to fire by 16 July. Technically, the date could be met, but the weather then became a question mark.[42] "Hence, there was feverish activity on our part," Hempelmann recalled, "to make the town monitoring program flexible enough to adapt itself to whichever wind condition prevailed when the test was ready."[43] Precautions against the danger of fallout over nearby towns suddenly became urgent.

Army intelligence agents crisscrossed the countryside trying to locate, list, and map every person living within a 40-mile radius of zero. Groves approved the formation of a 144-man army evacuation detachment under T. O. Palmer, Jr., a Corps of Engineers officer. Simplicity keynoted the plans. If fallout threatened only one or two families, Palmer's men would simply escort them to a hotel in town. Should a nearby town be menaced, Palmer commanded enough jeeps and trucks to remove as many as 450 people to safety; he also stocked enough tents, food, and other supplies to see them through two days. The distance of larger towns from zero allowed time to summon extra help if needed; Alamogordo Air Base prepared to furnish barracks for temporary housing. The army would merely explain that a large ammunition dump filled with gas shells had blown up and that the area must be cleared.[44]

Evacuation would occur only after monitors collected enough data to confirm real danger. Hoffman alone of those in the field might order the move. Such an order, however, would require an extreme emergency. Normally, the decision would belong to Stafford Warren, chief medical officer on Groves's staff, and his deputy, Hymer Friedell. Both planned to attend the test. Warren's post was Trinity base camp, in radio contact with Hoffman.[45] Friedell took an Albuquerque hotel room 100 miles away, so that "if something serious happened," he half-jokingly recalled, "at least it would only happen to half of us."[46] There he would await telephone reports from the field. Everything depended on the field reports from Hoffman's monitors. Limited in number, they needed a flexible plan to gear their movements to the observed course of the cloud. Hoffman decided to post them at the last possible moment. Only then would short-range weather forecasts grow firm enough to make him reasonably certain which way the wind was blowing.[47]

When Hoffman did make his final dispositions, the cloud appeared most likely to travel northeastward. He sent monitors to Carrizozo, Roswell, and Fort Sumner. From these New Mexico towns they could drive to highway points roughly 50, 90, and 150 miles northeast of zero to take readings and samples when the cloud passed. Even the day before the test, however, a chance the cloud might blow northwestward remained. Accordingly, Hoffman stationed a fourth monitor in Socorro, 35 miles to the northwest. Hoffman himself and four other monitors waited at the main gate, the northwest corner of the test site, ready to move either way the cloud moved.[48]

With help from Watts, Hoffman had also ringed the test site with recording meters and Filter Queen air samplers: in Carrizozo to the east, Tularosa to the southeast, Hot Springs (now Truth or Consequences) to the southwest, and San Antonio, Socorro, and Magdalena to the northwest. These were all towns between 30 and 55 miles from zero.[49] Hempelmann judged these arrangements adequate from a safety standpoint. Probably "the danger area will not extend beyond thirty miles," he informed Bainbridge; if the cloud drifted northwest, however, "our records of what happens beyond thirty miles will be unsatisfactory."[50]

Records were vital. Like shock and blast, fallout posed legal as well as safety problems. An army claims officer provided Hempelmann with detailed guidelines as to "what procedures should be used to legalize . . . monitoring records for the outlying country."[51] Hempelmann, in turn, passed them to Hoffman. The guidelines covered both recorded data and handwritten field notes. Army intelligence agents would witness the notes. Because fallout might also become a security problem, an agent rode with each monitor. Other agents waited in every town within 100 miles of zero. Ordered chiefly to observe public reactions, they intended to obtain legally valid records of bomb effects as well. Army intelligence also sent film packets by registered mail to all nearby post offices. Retrieved shortly after the test and processed under carefully defined conditions, they would provide another legal record. So would the analysis of earth samples collected by monitors.[52]

Evacuation, however, remained the most potent challenge to secrecy. Public safety might demand it, but only as a last resort. No one received advance warning; the army would move no one beforehand from the danger area. Only after the bomb exploded might the army act, and then only if the fallout cloud imminently threatened people in its path. But how serious would the threat have to become to trigger such action? In essence, safety planners had to decide how high an exposure to call safe for those with no part in, or knowledge of, Trinity.

Hempelmann and Nolan assured Bainbridge that a total dose of 68 roentgens spread over two weeks

> would certainly not result in permanent injury to a person with no previous exposure. . . . It would probably not even cause radiation sickness. A normal person could probably stand two or three times this amount without sustaining permanent bodily damage. Fatalities probably would not result unless ten or more times this dose were delivered.[53]

The question, however, remained open. In a meeting at Los Alamos on 10 July with Oppenheimer and his staff, Warren "took 60r in two weeks as safe. Even 100r would not be harmful provided there would be no further exposure." Much depended on the precise weather at the time, but Warren would begin to worry only "if peak reached 10r," that is, if the measured rate reached 10 roentgens per hour. The best approach, he suggested, was to take "measurements for several hours and consider evacuation if total dose reached final total of 60–100r."[54] At Trinity base camp two days before the test, Warren and Hempelmann agreed to "set the upper limit of integrated gamma ray dose for the entire body over a period of two weeks (336 hours) as 75 roentgens."[55] They also agreed on an "upper safe limit of radiation . . . [of] 15 r/hr at peak of curve."[56]

Despite his central role in safety planning for Trinity, Nolan departed before the test. He received orders on 1 July for a "special overseas assignment."[57] The army shipped Little Boy to Tinian Island in the Pacific for the first nuclear attack on Japan. It needed an escort from the Health Group who could also oversee safety measures on the island. Experienced and already in uniform, Nolan got the job. He remained at Trinity until 9 July, a week before the scheduled test date. When the test took place he was at sea aboard the USS *Indianapolis*. Hempelmann took charge of Trinity safety just as the physicists and engineers completed their preparations for shipping the gadget from Los Alamos to Alamogordo.[58]

FINAL PREPARATIONS

At seven o'clock Thursday morning, 12 July 1945, an armed convoy left Los Alamos for Trinity. Everyone carried pocket dosimeters. In the back seat of an army sedan rode a shielded and cushioned carrying case, proof against critical assembly of its contents in any road mishap. It held the gadget's plutonium core and three initiators. The core was a small, heavy sphere, thinly coated with nickel to protect it from oxidation. Coating also reduced human contact with the toxic metal. Workers who transferred core to case nonetheless wore masks and rubber gloves. Similarly protected workers at Trinity would insert the core into a uranium cylinder, precisely machined to plug a matching hole in the gadget's central uranium sphere. Concentric shells of polonium and beryllium formed an initiator, which would go into the core. When plutonium imploded, the crushed initiator would furnish neutrons to begin the chain reaction. Paul Aebersold oversaw the safe handling of the active metals. He noted that all pocket dosimeters read below tolerance for the day's exposure at the end of the journey.[59]

Thursday morning at Los Alamos another team began assembling the gadget. It comprised dozens of "explosive lens sectors shaped like watermelon plugs, all pointed toward the center."[60] They formed a hollow sphere measuring 5 feet across and weighing several tons. With assembly finished by midafternoon, workers loaded the cased, sealed, and waterproofed gadget aboard the truck. Quarter-inch steel cables secured it to a tub bolted to the truck bed. Thoroughly rehearsed the week before with a dummy gadget, the process went smoothly.[61] Meanwhile, a surprised Oppenheimer had learned that Hempelmann was still at Los Alamos working on laboratory contamination problems. He ordered the Health Group leader to Trinity at once, to oversee final planning for town monitoring.[62] Hempelmann joined the convoy, which left a few minutes after midnight on "Friday the Thirteenth, my choice, because I believed in unorthodox luck," said George B. Kistiakowsky, head of the Explosives Division.[63] Escorted fore and aft by cars full of armed guards, the gadget headed south. Although Trinity lies about 150 miles due south of Los Alamos, blacktop roads across rugged terrain allowed neither so direct an approach nor very high speed. The trip took all night and much of the morning. Before noon on Friday, however, the gadget rested on its cradle in the tent beneath the tower at zero.[64]

Hempelmann set off to confer with Hoffman.[65] Aebersold remained on the scene checking workers. Precautions included sharply restricted access to the area around the gadget. Everyone working there wore so-called catastrophe film badges designed to provide gamma readings from tolerance to several thousand roentgens and also including red phosphorus for neutron detection.[66] Catastrophe badges, Aebersold explained, were "intended to measure personnel exposure in case of a partial premature detonation."[67] As he observed, however, "no reading above background was found on the badge of any member of the assembly crew." Portable alpha counters likewise detected no trace of contamination on any worker.[68]

Initiator and core reached the tower at 3:18 Friday afternoon. Held at a deserted ranch house 2 miles away, they had been mated just before the trip to zero. Tension ran high. For the first time, high explosives would surround an active core. The cradled gadget lacked a single explosive lens. Through this gap the uranium plug holding the active core had to be inserted into the central uranium sphere.

> Imagine our consternation when, as we started to assemble the plug in the hole, deep down in the center of the high explosive shell, it would not enter! Dismayed, we halted our efforts in order not to damage the pieces, and stopped to think about it. . . . Meantime, a desert sandstorm had come up, accompanied by the usual midafternoon dry thunderstorm. It was indeed a frightening situation. We were located under a 100 foot steel tower, which protruded like a lightning rod above the plain, while the wind and lightning were playing around the structure.

Pausing for thought while plug touched sphere proved vital. Afternoon heat in the ranch house and the hot car ride to the tower had expanded the plug, just enough to exceed the few thousandths of an inch clearance between plug and hole. Contact balanced the temperatures. "Finally, then, the plug was lowered gently home." For the rest, things went smoothly and the gadget awaited the next step.[69]

Work resumed at eight o'clock Saturday morning. After striking the tent, the crew hoisted the bomb to the sheet-steel shed atop the tower. Detonators and firing circuit completed the device. All was ready by five o'clock that evening. Just before midnight, final rehearsal began. Earlier in the week, two trials—rescheduled for mornings to avoid the daily afternoon thunderstorm that lent a touch of drama to the Friday scene at the tower—had centered on instruments and circuits. The final exercise, however, also included camp guards, Health Group monitors, and army intelligence agents.[70]

> Clearance of the area, reporting to shelters, and even post-shot monitoring operations were carried out as if it were the actual test. At zero-time rehearsal night a small charge was even fired (at some dis-

tance from the tower) to test the arming and firing devices and to signal the monitors to proceed.[71]

The only task scheduled for Sunday was looking "for rabbit's feet and four-leaved clovers."[72]

Debilitating heat and the harsh desert setting compounded the mounting tensions of the final week. Nothing, however, slowed steady progress toward Trinity's planned climax at four o'clock Monday morning, 16 July 1945. Doubts lingered about how well the bomb would work. More pragmatically, planners wondered if the last instances of electrical interference between experiments and instruments were truly resolved. But weather remained the big question on Trinity eve. Stafford Warren had arrived at Los Alamos in the final week. As chief medical officer for the Manhattan District, he had campaigned hard at headquarters for safeguards against fallout. He would also play a key role in any decision about evacuation or other emergency measures. "Consultant," however, was his formal title for Trinity. He spent the week in almost continuous meetings with Hempelmann and the Health Group about their plans, and about the weather.[73]

During the last weekend Warren and Hempelmann reviewed final safety plans, not only for workers but for the host of dignitaries and onlookers gathered for the big show. One-page summaries of safety rules were posted around the base camp. They were also read and discussed in group meetings during the final hours. Only those with jobs to do remained at test shelters, or as near as the base camp. Others joined the group forming on the height 20 miles northwest of zero, named Campana Hill for the occasion. The rules were simple: wear pants and long sleeves to protect skin from ultraviolet light; lie prone with feet toward zero and eyes covered to avoid blast and light damage; look at the fireball only through welder's dark glass. A siren would announce time to lie down and would later signal all clear. Until the last moment at both base camp and Campana Hill, public address systems periodically repeated the rules. The same systems would direct evacuation if it came to that.[74]

These final precautions applied to those who knew what to expect and could be told what to do. Beyond Trinity's borders, people knew nothing and could be told nothing. Their safety depended on Hoffman's monitors and Palmer's evacuation unit. Perhaps even more, it depended on the vagaries of the weather. When Hoffman left Los Alamos for Trinity late Saturday afternoon, the Monday forecast predicted light and variable winds. He met his teams in Santa Fe Sunday afternoon. Monitors and intelligence agents received their instruments and final instructions, then headed for their stations. Warren's base-camp radio or Friedell's Albuquerque telephone would tell them how high, how fast, and which way the cloud was moving after the blast.[75]

The more distant monitors stationed to the northeast could wait; hours would pass before the cloud reached them, its activity by then much reduced. Their chief concern was valid records. For Hoffman, his two crews, and Palmer's men poised at Trinity's main gate, quick data were more urgent. If the cloud blew northwest, a ranch only 15 miles from zero and a town 5 miles beyond might require speedy evacuation. Hoffman and Palmer had three radios between them to receive cloud data as swiftly as possible.[76]

Sunday night Bush's military police fanned out along the access roads from zero, making certain no one lingered in the test area. After they reported all clear, Bush double-checked the rosters to confirm that everybody had left. Just before eleven o'clock he met Bainbridge's arming party at the base camp. Three cars drove toward zero in a misting rain. The weather had taken an unexpected turn for the worse. Unpredictable winds would not delay the test. The problems they might cause had, in any event, been foreseen and steps had been taken to cope with them. Rain posed a far graver threat. Fallout washed from the air over a restricted region could create dangerously high exposure levels on the ground. Whether or not the gadget would be fired Monday morning suddenly became a real question.[77]

Groves had arrived at Trinity only that evening, having deliberately waited until the last minute to avoid any impact his presence might have on test preparations. He disliked the scene that greeted him:

> To my distress, I found an air of excitement in the base camp instead of the calmness essential to sound decision-making. There were just too many experts giving advice to Oppenheimer about what he should do with the majority of them advising postponement. And, what was worse, none of these were experts in the area that mattered. Was it going to rain and how much? When, and for how long?

Groves strongly opposed anything but a temporary delay. The test was still hours away, and the weathermen "saw no reason for not waiting until the last minute to decide."[78]

Postponement might compromise President Truman at Potsdam and threaten the schedule for bombing Japan. Worse, it might not even be the safer choice. Project morale would surely suffer. Pragmatically, the longer the delay, the more likely a malfunction in the miles of cable and wiring which could cause the bomb to misfire. A lengthy delay might produce just the disaster so much effort had gone into preventing. As Groves explained, "We simply could not adequately protect either our own people or the surrounding community or our security if a delayed firing did occur."[79] For Bainbridge, that fear lay closer to home: "My personal nightmare was knowing that if the bomb didn't go off or hang-fired, I, as head of the test, would have to go to the tower first and seek to find out what had gone wrong."[80]

Four men shared the decision to fire: Oppenheimer, Bainbridge, Hubbard, and General Thomas F. Farrell, Groves's deputy. They planned to meet at 2:00 A.M. to cast the final vote for the 4:00 A.M. shot, having agreed that the vote must be unanimous. At two o'clock, however, Bainbridge was still waiting at the tower for a better weather report. Hubbard, who favored going ahead, "assures everyone there is no possibility of rain following the shot."[81] That was hard to believe at the tower. "At about two o'clock in the morning" Boyce McDaniel recalls climbing the open ladder for a final check of the gadget and seeing "thunderstorms playing around the site with frequent flashes of lightning followed by rolling thunder."[82] Local rain had stopped, but the sky remained heavily overcast. The test was postponed for an hour. Bainbridge and his arming party huddled in their cars near the foot of the tower. Occasionally, one dashed to the phone on the tower's leg for the latest weather report.[83]

Finally, the overcast began to break up. Just before 4:45, Hubbard told Bainbridge by phone that the weather looked good enough to fire at 5:30—not ideal, but good enough. Bainbridge called Oppenheimer and Farrell. They all agreed to a 5:30 deadline. After throwing the last switches to arm the bomb and allow its firing from S 10,000, the arming party retreated. "The drive out to S 10,000 was not made above 35 mph, contrary to rumors."[84] At the test shelter, the decision came over the loudspeaker at 5:10. Zero hour would be 5:30. The final countdown began. Announcements of the time over loudspeakers punctuated the waning moments. It was a very long twenty minutes, the last seconds longest of all.[85]

THE DETONATION

In his formal account of Trinity, Bainbridge limited his comments on the shot itself to a dry recital of the figures:

The location of Point 0 was latitude 33°40'31", longitude 106°28'29", based on New Mexico Map No. 4, Grazing Service, Albuquerque Drafting Office. The time is known only very poorly because of difficulty in picking up station WWV for a time check. The best figure is July 16, 5:29:15 A.M. MWT [Mountain War Time] plus 20 s [seconds] minus 5 s error spread.[86]

Like so many others, he later tried to capture the human meaning of the event:

I felt the heat on the back of my neck, disturbingly warm. Much more light was emitted by the bomb than predicted. . . . When the reflected flare died down, I looked at Oscuro Peak which was nearer Zero. When the reflected light diminished there I looked directly at the ball of fire through the goggles. Finally I could remove the goggles and watch the ball of fire rise rapidly. It was surrounded by a huge cloud of transparent purplish air produced in part by the radiations from the bomb and its fission products. No one who saw it could forget it, a foul and awesome display.[87]

The test was a great success. The bomb seemed a good deal more powerful than predicted.[88] If it prompted Bainbridge to remark, "Now we are all sons of bitches!"[89] or reminded Oppenheimer of a passage from the *Bhagavadgita*—"I am become death, the shatterer of worlds"[90]—the more typical response echoed Kenneth Grieson's "My God, it worked!"[91] Within minutes Groves was on the phone to Washington, clearing the way for President Truman's Potsdam ultimatum. Concern now focused on fallout. Immediate hazards were as modest as expected for those on the test site. Aebersold's safety plan called for the men stationed at the three test shelters to take readings and leave for the rear area. They had half an hour before the cloud reached them. W 10,000 followed the plan. Departure from N 10,000, however, was hasty.[92]

As expected, no meters in or around the shelters showed anything at first. Initial radiation from fission and other processes in the explosion ceased in less than a minute; delayed neutrons lasted only seconds. Radiation from the fireball, although substantial, decreased as the square of the distance, further attenuated by the air. Ten thousand yards from zero it was simply too low to detect. But as the cloud drifted northward, remote sentinels along the access road from zero transmitted rising readings to N 10,000. Then Henry L. Barnett, the Medical Corps captain serving as chief monitor at the shelter, also observed a large and rapid increase on his Watts-type meter. He requested and received permission to evacuate the shelter. Barnett's reading proved a false alarm, the result of a faulty zero setting on his meter. As Aebersold reconstructed the event, a puff of active material had drifted by; sentinel readings were real enough, and he found 0.01–0.02 roentgen per hour at the shelter when he checked just over two hours later. That was higher than readings at the other shelters, but not seriously so. Film-badge readings later confirmed that no one at N 10,000 had been exposed to more than 0.1 roentgen.[93]

S 10,000 evacuated only in part. Unexpectedly efficient, the explosion and thermal updraft sent the fireball high enough to forestall much wind-borne fallout closer than 10 miles from zero. So it appeared, at any rate, to Trinity analysts; Aebersold's sketchy postshot survey in fact found no significant activity anywhere on the test site, except within 1,200 yards of zero. Elsewhere the highest on-site gamma readings fell below 0.1 roentgen per hour.[94] Contamination by fission products or plutonium ranged from "far below tolerance values" to nothing.[95] An inversion layer at 17,000 feet did trap some radioactivity. Later that morning the rising sun warmed the valley air. Turbulence carried cooler air from above with its burden of fission products past S 10,000. No one was too concerned, once Bainbridge had made some hasty calculations. "Hempelmann's counters began to click more and more rapidly. . . . We put on dust masks as the counting rate increased. Finally the radioactivity decreased as the air was slowly swept away."[96]

Radioactivity in the crater far exceeded pretest estimates. Military police, guided by Aebersold's preliminary survey and the red flags posted before the shot, set up temporary roadblocks to control access to the crater. First in were Herbert Anderson's earth-sampling tanks. They left S 10,000 at seven o'clock, an hour and a half after the shot. Pausing at the sentinel 1,500 yards south of zero, they found readings less than 0.1 roentgen per hour. One tank then crawled toward the edge of the crater. The tower was gone, vaporized in the blast. Peering through his periscope, Anderson saw only a shallow bowl crusted with green glass, 10 feet deep and perhaps 400 yards across. Surprisingly high activity compelled Anderson to grab his samples and quickly depart.[97]

The lead-lined tank visited the crater four more times that day and the next, once on the second day passing through its center. Activity remained very high, "far beyond the range of our measuring instruments (even with the 50 fold shielding factor provided by the tank)," was the later report; "highest radiation intensities could only be estimated [at] approximately 600–700 roentgens per hour."[98] Higher than expected crater readings, Anderson later explained, resulted from a simple oversight. He and his colleagues had failed to allow fully for the effects of neutron-induced activity in sodium and other soil elements.[99] Pocket-dosimeter and film-badge readings showed that one army tank driver accumulated between 13 and 15 roentgens in three trips; the other, 3.3 roentgens in two. Anderson's total fell between 7 and 10 roentgens for two trips. A second two-time observer accumulated 7.5 roentgens. The third observer received 5 roentgens during his single ride.[100]

The rest of Anderson's team stayed farther from the crater. Most went no nearer than the 1,500-yard sentinel. The highest film-badge exposure recorded for any of them at that distance was 0.26 roentgen. Several, however, briefly ventured 1,000 yards closer in the second tank, equipped with the rocket sampler but unshielded. The highest film-badge reading for any member of this group was 1.0 roentgen. All Anderson's men wore protective clothing and respirators. Checked at S 10,000 by Health Group monitors after they returned from the crater, their contamination seemed largely limited to outer clothing. This removed, they showered to dispose of what little active dust clung to heads, hands, or feet. The glassy crust over the most active region made contamination by dust a notably smaller problem than had been expected.[101]

Only three other entries into the flagged region around zero occurred on the day of the shot. Richard Watts of the Health Group escorted a photographer to Jumbo, the unused steel vessel designed to contain a fizzled explosion. Protectively clothed and masked, they remained 800 yards northwest of zero from eleven until noon. The measured exposure rate at 11:30 was 1.0 roentgen per hour; each man received an estimated 0.5 to 1.0 roentgen. Two parties of physicists in full protective garb entered the area at 2:30 that afternoon. Three men went to a point 800 yards south of zero to retrieve neutron detectors. The task took half an hour and produced their single film-badge reading of 0.1 roentgen. Three others, also on a data-recovery mission, drove a jeep near the edge of the crater, 350 yards southwest of zero, where radiation intensity was 6 to 10 roentgens per hour. As they stayed only ten minutes, the highest film-badge reading was 0.86 roentgen.[102]

All other entries were postponed, with several factors influencing this change of plans. One was the higher than expected activity in the crater region. Recovery of experimental data, in any event, seemed less urgent after the test's obvious success. Perhaps most crucial, though, was the diversion of Health Group monitors from the task of clearing going-in parties through S 10,000. They found themselves increasingly preoccupied with fallout problems beyond the test site. Initially, within minutes after the shot, the cloud had formed a dense white mushroom 25,000 feet from top to bottom above a reddish brown stem 10,000 feet high. An hour later the cloud was largely lost to sight, spreading out and fading indistinguishably into the other thin white clouds sharing the morning sky.[103] What could not be seen, however, could still be detected. Rising gamma readings briefly marked the passage of the cloud overhead, "dropping its trail of fission products."[104] That longer-lasting track on the ground could be followed by other instruments.

FALLOUT

Light to gentle winds blew mostly toward the northwest near the ground but northeastward aloft. Monitors detected the first traces of fallout, in fact, toward the northwest. By 7:00 A.M., however, the critical region clearly extended north and northeast. Hoffman dispatched monitors to highway 380, paralleling the test site's northern border. They had orders to survey the highway eastward toward Bingham. Bingham, like most other places that figured in the day's events, was little more than a country crossroads store.[105]

Hirschfelder and Magee drove up in the battered old sedan they had borrowed from a Los Alamos colleague for the occasion.

> John and I rang the door bell and an old man came out. He looked quizzically at us (John and I were wearing white coveralls with gas masks hanging from our necks). Then he laughed and said, "You boys must have been up to something this morning. The sun came up in the west and went on down again." There was some fallout there, but the level of radioactivity was not dangerous. The soldiers bought almost everything in the store and left his shelves bare.[106]

Hoffman soon took the road to Bingham himself, joined by Palmer and some of his troops. The next few hours were hard ones for the monitors and anxious ones for Groves and his lieutenants at the base camp. Vagaries of wind and terrain produced anomalies in the fallout pattern, a problem compounded by lack of radio links between monitors. They could trade information only when their paths crossed or at prearranged meetings.[107]

Hirschfelder and Magee moved past Bingham to an army searchlight post, set up to illuminate the cloud in the predawn darkness.

> The soldiers there had bought some huge T-bone steaks to eat after the atom bomb explosion. When we arrived they were just roasting the meat and it smelled delicious. However, at the same time the fallout arrived—small flaky dust particles gently settling on the ground. The radiation level was quite high.[108]

Magee read 2.0 roentgens per hour at 8:30 A.M. The handwritten scrawl of one of the searchlight crew members told the story: "buried steaks which were cooking on open fire because they tho[ugh]t they were too much contaminated & so buried them & pulled out."[109]

Hoffman's radio contact with Warren at the base camp, poor to start with, grew worse; nobody had thought to check radio ranges in advance. Just after nine o'clock Hoffman reported Magee's reading of 15 roentgens per hour on the county road 7 miles northeast of Bingham. This alarming news was the last message he was able to transmit. Warren summoned Hempelmann from S 10,000 to confer. They sent two of Aebersold's monitors to help Hoffman. Hempelmann located a car equipped with a good radio and followed them. Meanwhile, Hoffman found levels as high as 3.3 roentgens per hour at Bingham, Adobe, and White along highway 380. Such readings suggested that total exposure in that area might approach the allowed limit. Assuming the worst, Hoffman projected an integrated gamma dosage at 90 percent of tolerance. Unable to reach Warren by radio, he dispatched a courier with the message.[110]

Hempelmann found Hoffman at Bingham just before noon. Most of the other monitors were also there. The picture had become both clearer and less alarming as the dust settled—literally. Readings dropped sharply in most places when fallout reached the ground and airborne activity decreased. Remaining concern focused on the anomalously high readings from the stretch of county road climbing toward Chupadera Mesa northeast of Bingham. The conference was brief. One monitor drove back along highway 380 to make certain the situation there remained under control. Two others headed north, circling the anomalous region and trying to locate its northern limits. Hoffman and the rest of the monitors made for the point of highest readings, taking along some of Palmer's troops against the possibility of evacuation.

Palmer and the bulk of his command returned to their bivouac on the test site. Hempelmann radioed Warren an account of the plans, then drove back to the base camp himself.[111]

Hoffman confirmed high readings northeast of Bingham. The highest appeared where the road ran through a steep gorge, accordingly nick-named the "Hot Canyon." As his maps showed no nearby dwellings, no action seemed required. The day's last flurry of concern came late that afternoon. Winds shifting to the east blew active dust back toward Carrizozo, the town nearest the eastern border of Trinity. There two physicists, husband and wife, waited in a tourist court cabin to obtain seismograph and radiation records. They phoned to report the Geiger counter going off scale. This counter, however, was their more sensitive instrument. When the level rose high enough for the Victoreen meter, it proved no more than 0.0015 roentgen per hour. It also quickly de-cayed; almost no trace remained when they left the next morning. Car-rizozo was the only town of any size even momentarily considered for evacuation. By Monday evening, in fact, despite still sketchy returns, everyone believed that Trinity fallout had caused no serious danger anywhere.[112]

Tuesday, however, produced a surprise. Puzzled by the Hot Canyon, Hempelmann and Friedell drove out for a closer look. They discovered an adobe house hidden from the road a mile east of the highest read-ings. Somehow overlooked by the army, it had been omitted from the monitors' maps. An elderly couple named Raitliff lived there with their young grandson, several dogs, and assorted livestock. Radiation levels in the locale, the doctors decided, were not high enough to demand hasty evacuation. When all the monitors met two days later at Los Alamos, they discussed the question of the Raitliffs. The decision against evacuation stood. Regular monitoring of the area began at once, however, and persisted over the next six months. "Visits were made periodically to the families in the most heavily contaminated regions to determine whether their health had been affected."[113]

A second ranch unknown to the army came to light later. A couple named Wilson lived near the Raitliffs, and early reports confused the two. According to Hempelmann's figures, the Raitliffs received at most a total of 47.0 roentgens of whole-body gamma radiation in the two weeks after Trinity. He put the twelve-week total at 49.4 roentgens. These were estimates calculated on the basis of scattered nearby read-ings and known rates of fission-product decay, with some allowance for the shielding provided by adobe walls. Monitors found radiation levels near the Wilsons roughly three-fourths the Raitliff readings. Once again, they judged no action required. Fallout at places like White and Bingham produced much lower readings and aroused little concern.[114]

Hempelmann visited the Raitliff ranch again in mid-August. He learned that the ten-year-old grandson had left home early in the morning of 16 July to spend the day at Bingham. Since he also remained indoors that night, Hempelmann concluded that "he missed most of the heavy exposure of the first day in the 'Hot Canyon.' " The house, a well-built adobe structure with walls 15 inches thick, "affords great protection," as indoor readings were a tenth those outside. Thus Mrs. Raitliff also escaped heavy exposure, since her work kept her inside most of the day.[115] Only Mr. Raitliff had spent much time in the open that day; "the ground immediately after the shot," he told Hempelmann, appeared "covered with light snow." Especially at dawn and dusk, he added, for several days "the ground and fence posts had the appearance . . . of being 'frosted.' "[116] Hempelmann, however, judged his health good. Injuries seemed limited to some animals with beta burns.[117]

Hempelmann's last visit to the Raitliffs came three months later. The grandson had left in September to stay with other relatives nearer school. The Raitliffs, however, were now caring for a two-year-old niece. Excepting colds, they all looked in good health to Hempelmann. The animals, he noted, no longer displayed the burns, bleeding, and loss of hair he had observed earlier. Signs of damage, however, persisted. One bitch still limped, although her paws showed none of the rawness and bleeding—the apparent aftermath of beta burns—seen on the August visit. Patchy and discolored coats also marked a number of the animals, cows as well as dogs. The afflicted animals had all been in the open when fallout descended; active particles had apparently sifted through their coats. Particles lodged against the skin irradiated and damaged the tissue they contacted. The dogs' injured feet could be blamed on particles caught between their toes, since none of the hoofed animals suffered such damage.[118]

Animals, in fact, seemed Trinity's chief victims. The high grasslands around the site, particularly on Chupadera Mesa, grazed many cattle and fallout dusted some. Army officers who surveyed the region after one rancher filed a damage suit in October concluded that as many as six hundred cattle were affected. None seemed badly injured—superficial local burns and temporary loss of dorsal hair—but patches of hair grew back discolored and the animals' market value suffered. The army offered to buy as many as the ranchers agreed to sell, seventy-five head in all.[119] Los Alamos received the seventeen animals most obviously marked; the rest were shipped to Oak Ridge, where their descendents still dwell. Ultimately, no doubt remained that the damage resulted from Trinity fallout.[120] Robert Stone, former head of the Chicago Health Division, saw some of the cattle when he visited Los Alamos in February 1946. Judging from wartime experiments, he told Hempelmann, the cause of damage looked like beta radiation. He estimated "the dose . . . required to produce such an effect . . . to be between 4,000 roentgens and 50,000 roentgens, probably about 20,000 roentgens."[121]

Injured cattle, however, signaled no danger to people. As Wright Langham observed, "Cattle don't bathe." Active dust reaching their skin stayed there, "constantly decaying and irradiating the skin of the animals." People, of course, also wear clothes, making any chance of such skin damage even more remote.[122] Ill-informed members of the public, however, might see the matter in another light. Accordingly, Oppenheimer held Health Group reports on Trinity even more tightly than normal. Segregated from other Trinity reports, they were released only at Oppenheimer's personal request. He wanted, so at least one of his colleagues believed, "to safeguard the project against being sued by people claiming to have been damaged."[123] Exposure to the whole body from the fallout levels measured, in any event, appeared well below anything that might cause harm, save perhaps to the families in the Hot Canyon. Observations of the families and their surroundings over the next few months, however, allayed the worst fears. Elsewhere, the highest readings on the ground outside the test site nowhere exceeded the 3.3 roentgens per hour Hoffman found in Adobe, although readings twice as high briefly marked the cloud's passage.[124]

Forty-four monitors, scientists, and army intelligence agents had been posted north and east of zero. They easily tracked the cloud across New Mexico and into Colorado. Fallout levels higher than 0.5 roentgen per hour, however, covered a relatively narrow region, an irregular swath thirty miles at its widest, stretching a hundred miles northeast of zero. The pressure of other wartime concerns precluded any detailed fallout survey at the time.[125] In mid-December 1945 two monitors took a closer look at the area along and north of highway

380. After three days and 377 miles, they reported their findings. Activity levels had, of course, fallen. The Hot Canyon, however, still deserved its name. The 19 milliroentgens per hour of gamma radiation they detected near the Raitliff ranch was nearly double the highest reading anywhere else.[126] A full-scale survey, however, waited until three years after Trinity. To conduct it, the new Atomic Energy Commission contracted with the University of California, Los Angeles, where Stafford Warren had become dean of the medical school. Other than the crater region, the UCLA team found the largest residues, still easily measurable, on Chupadera Mesa, grazing range of most of the injured cattle.[127]

Trinity camp itself quickly settled into routine. Most test workers left for Los Alamos the same day. Heavy demand for clearance to retrieve instruments and data abated within another two days. For several weeks the Health Group retained one or two officers on the spot to check parties into the crater region, but after 11 August it simply provided information, equipment, and help if requested. Military police on foot and horseback patrolled the fences around the crater. They also maintained a roster of all entrants. Although MPs checked most passes, they themselves required no clearance and often asked none of other soldiers stationed at Trinity.[128]

Not until October, during a chat with the MP commander, did Hempelmann realize what was happening. Several MPs regularly escorted parties into the crater, and some of the soldiers made frequent visits. Trinity had become something of a mecca for sightseers. The practice was immediately halted. After a detailed study of the records and interviews with the soldiers, Aebersold concluded that two of them might have received as much as 30 roentgens between mid-August and mid-October. No one, however, showed any sign of harm, and the likely totals fell well below these maximum estimates.[129]

Whatever questions might later surface, on 16 July 1945 nothing really mattered beyond the fact that the implosion bomb had worked and had worked better than all but the most optimistic planners predicted. Fortunately, Trinity's isolation in a sparsely peopled region apparently achieved it purpose: no one hurt, the secret safe. The weather-caused delay meant more people knew something happened than Groves preferred, for firing on schedule would have found more of them sleeping. Flash, noise, and some broken windows brought questions from a few local reporters. They heard about a harmless accident

in a remote ammunition dump. No large newspaper picked up the
story. The only anxious moments came from delays in learning the fate
of the cloud. As Groves remarked, however, "within a few hours, we
knew that we were in the clear."[130] Six weeks later Hempelmann assured him that "radioactive materials did not fall from the cloud outside
the [crater] area in amounts which would be considered dangerous."[131]

Despite injured cattle, no evidence suggested that any person suffered exposure to what were then believed to be harmful amounts of
radiation. Still, as Warren observed, in some respects it was a near
thing:

> While no house area investigated received a dangerous amount, i.e.,
> no more than an accumulated two weeks dosage of 60r, the dust out-
> fall from the various portions of the cloud was potentially a very dan-
> gerous hazard over a band almost 30 miles wide extending almost 90
> miles northeast of the site.

Alamogordo, he concluded, "is too small for a repetition of a similar
test of this magnitude except under very special conditions." He proposed finding a larger site, "preferably with a radius of at least 150
miles without population" for any future test.[132]

5

FROM JAPAN TO BIKINI

HIROSHIMA AND NAGASAKI

With Trinity, Los Alamos completed its central task of developing and proving fission bombs. Delivering combat-ready bombs to the Army Air Force remained the final step. Explosives expert and career naval officer William S. Parsons, Los Alamos Ordnance Division leader since 1943, took charge of this effort in March 1945. He left for the Pacific after Trinity to direct final assembly of Little Boy, the untested but surefire uranium-gun bomb. Bad weather over Japan delayed the first nuclear attack until 6 August, but that morning Little Boy exploded 1900 feet above Hiroshima, low enough to maximize blast effects, planners believed, but high enough to keep the fireball from touching the ground and thus to avoid severe radioactive contamination.[1]

Parsons, the only Trinity witness aboard the bomber, radioed the first report fifteen minutes after the drop: "Results clear-cut, successful in all respects. Visible results greater than Trinity."[2] Appearances deceived Parsons: the estimated yield of Little Boy fell well short of Trinity's, 13 versus 19 kilotons. The plutonium-implosion bomb nicknamed Fat Man devastated Nagasaki three days later. Detonated like Little Boy high above the city, it had an estimated yield much higher, 23 kilotons.[3] Technically, the bombs worked. Politically, they seemed to work as well. Preparations for a third attack were suspended, then canceled. Six days after Nagasaki the war ended, although the formal signing of surrender papers did not take place until 2 September 1945.[4]

Japanese authorities meanwhile found themselves ill equipped to cope with the disaster. The injured and dying numbered in the tens of thousands. Devastation wreaked by the bombs further hampered rescue and relief work in a country already much weakened by widespread shortages of food, medical supplies, and other basic needs.

Such efforts nonetheless began at once. So, too, did attempts to learn the precise nature and scope of the disaster; Japan received no explicit warning of the unique hazards the bombs might pose.[5] Japanese teams inspected casualties, performed autopsies, collected soil samples, and measured radioactivity even before hostilities ended on 15 August. With the war over, the scope of both aid and study expanded. The Science Research Council of Japan formed the Special Committee for the Investigation of A-Bomb Damages on 14 September. Its purpose was to collect, compile, and publish systematic data on bomb effects. By then, the first American survey teams had already arrived.[6]

The Manhattan Project planned no postaction survey of bomb effects until Tokyo radio reported mysterious deaths among relief workers at Hiroshima and an American scientist, once briefly employed on the project, made headlines by claiming Hiroshima might remain uninhabitable for seventy years. Japan's formal charge on 10 August that atomic bombs violated the laws of war may have tipped the balance. Radioactivity had clearly become an issue.[7] Manhattan Project commander Leslie Groves ordered his deputy, Thomas Farrell, to go to Japan. Already at Tinian, forward base for the nuclear attack on Japan, Farrell at once began forming the Manhattan Project Atomic Bomb Investigating Group. Physicists and technicians on the island provided the core; Tinian had few doctors, however, and Farrell had to await a medical team from the United States. Organizing that team fell to Stafford Warren, Grove's chief medical officer. Uniformed members of the Los Alamos Health Group filled most of the slots.[8]

Groves wanted quick surveys of Hiroshima and Nagasaki. The main reason offered for haste was the imminent entry of Allied troops, a point stressed in the message from Washington asking General MacArthur to expedite the survey. Speed was vital to assure "that these troops shall not be subjected to any possible toxic effects, although we have no reason to believe that any such effects actually exist."[9] Perhaps more important, though, Groves wanted proof that radioactivity had played no major role in the overall damage. Donald Collins, a health physicist with the medical team, recalled Farrell's blunt statement that "our mission was to prove that there was no radioactivity from the bomb."[10] Warren's orders called for rapid surveys of the bombed cities to forestall any question of troop safety. They also required him to provide records of any residual ground contamination and to report the full extent of blast and other bomb-caused damage.[11]

Warren joined Farrell in mid-August. Island-hopping from Tinian through Guam and Okinawa, they reached Tokyo early in September. To their surprise they found another medical team already there. Uncertain about Manhattan Project plans, MacArthur's headquarters formed its own group late in August. The moving force was Ashley W. Oughterson, a Yale professor appointed colonel in the Army Medical Corps and in 1945 serving as chief consulting surgeon in the western Pacific. Scouring the command for likely recruits, Oughterson found few radiologists but a dozen doctors with some research background. He led this team into Tokyo on 1 September, the day before Japan's formal surrender. The team's members intended chiefly to study medical and biological effects of the bombings. Complementary goals induced Warren and Oughterson to pool their resources: Warren, under severe time pressure, welcomed aid from Oughterson's doctors in his radiation and bomb-damage survey; for his more thorough studies, Oughterson could make good use of any data Warren collected.[12]

Warren's party reached Hiroshima on 8 September, transported in six cargo planes loaded with emergency supplies.[13] Farrell described the scene on the ground as "awe-inspiring and tremendous. A city of approximately 300,000 was essentially destroyed. While there were many buildings standing around the outer part of the city, its center was leveled."[14] The survey began in that leveled center. Carrying Geiger counters and Lauritsen electroscopes, members of the team worked their way to the circumference, taking readings as they went.[15] The first quick look at Nagasaki came later that week. The second city presented a different picture, showing "more spectacular effects," Farrell thought. "The blast was of much greater power . . . , twice that at Hiroshima." Nonetheless, he noted, "the rugged terrain, including steep hills and deep ravines, provided much shielding. . . . Because of that fact, Nagasaki is still alive and functioning, while Hiroshima is flat and dead."[16] With preliminary surveys completed on 14 September, Farrell returned to Tokyo for a plane home the next day. The two-day surveys of Hiroshima and Nagasaki gave him what he needed. In Hiroshima, he reported to Groves, the survey found no trace of radioactivity "on the ground, streets, ashes, or other materials." Nagasaki likewise "disclosed . . . no evidence . . . of any radioactivity in . . . samples of dirt, wood, and metal from all over the blast area."[17] Farrell may have been too sanguine.

Most of Warren's team remained in Japan to survey the bombed cities more thoroughly; even early September's quick look had detected some activity in both. The later surveys in Nagasaki, 20 September to 6 October, and Hiroshima, 3 to 7 October, clarified the picture. Much the same pattern was found in both cities: an active area centered beneath the point of detonation, the hypocenter; and a second, more diffuse area some distance downwind. Residual activity at the hypocenters fell below 0.1 milliroentgen per hour, fading away smoothly in all directions. Subsequent analyses found that such a circular pattern was most likely to have been caused by neutron-activated soil elements. Radioactive areas downwind, on the other hand, seemed more likely owing to residual fission products. Rain fell on both cities within an hour after detonation, washing fission products from the air. This so-called rain-out measured up to 0.05 milliroentgen per hour slightly west of Hiroshima, two or three times natural background levels. Measurements up to 1.0 milliroentgen per hour marked an area east of the Nagasaki hypocenter, near Nishiyama reservoir.[18]

The findings left much room for judgment. "The instruments," Warren observed, "functioned satisfactorily." Yet readings, even under the best conditions, might vary as much as 30 percent. Conditions, particularly in Hiroshima, were far from the best. Obliterated roads, heavy rains, and limited time made for "a somewhat less [than] complete survey." Radioactivity in all areas checked, Warren nonetheless concluded, fell "below the hazardous limit; when the readings were extrapolated back to zero hour, the levels were not considered to be of great significance."[19] In its report, the Manhattan Project Atomic Bomb Investigation Group explained what that meant:

> These calculations showed that the highest dosage which would have been received from persistent radioactivity at Hiroshima was between 6 and 25 roentgens of gamma radiation; the highest in the Nagasaki Area was between 30 and 110 roentgens of gamma radiation. The latter figure does not refer to the city itself, but to a localized area in the Nishiyama District. In interpreting these findings it must be understood that to get these dosages, one would have had to remain at the point of highest radioactivity for 6 weeks continuously, from the first hour after the bombing. It is apparent therefore that insofar as could be determined at Hiroshima and Nagasaki, the residual radiation alone could not have been detrimental to the health of persons entering and living in the bombed areas after the explosion.[20]

Such calculations posed serious problems and scarcely addressed the crucial matter of dose. "I don't think we could be sure of the order of magnitude of exposure to a given patient," Collins recalled.[21] Estimating doses, however, was not the main purpose of the Manhattan Project survey. First came troop safety. Radioactivity appeared nowhere to approach levels high enough to expose measurably, much less threaten, occupation forces. Most damage, to people as well as to property, derived from heat and blast effects. Radiation injuries in both cities seemed to Warren solely the result of the initial burst of gamma rays and neutrons emitted when the bombs exploded.[22]

ASSESSING THE DAMAGE

Testifying before Congress after his return, Warren claimed that radiation caused less than 8 percent of the deaths.[23] Most experts believed this figure far too low. One public challenge came from the United States Strategic Bombing Survey. Established in 1944, the survey assessed the results of air attacks on Germany as a guide to bombing Japan. When the Pacific war ended, President Truman ordered the survey to Japan. A four-month study by a team of 1,150 began in September. It covered the entire bombing campaign, with special studies of Hiroshima and Nagasaki. Radiation effects, the survey concluded, caused up to 20 percent of the deaths. The figure could easily have been much higher. Collapsing buildings, fire, and other causes killed outright many persons exposed to levels of initial radiation which in themselves would have proved lethal in days or weeks.[24]

Exactly how many died as a direct result of the bombings remains unclear. Estimates at the time varied widely, for a number of reasons: uncertain population figures, disrupted record keeping, delayed effects. Various groups placed the total killed and injured "between 100,000 and 180,000 for Hiroshima, and between 50,000 and 100,000 for Nagasaki," noted the Strategic Bombing Survey. The survey itself concluded that the Hiroshima dead totaled "between 70,000 and 80,000, with an equal number injured; at Nagasaki over 35,000 dead and somewhat more than that injured seem the most plausible estimate."[25] More recent efforts to fix the death toll have sometimes gone much higher: 140,000 in Hiroshima by the end of 1945, 70,000 in Nagasaki. Others have proposed even higher figures.[26]

Questions of long-term effects have become perhaps even more problematic; the absence of any direct records of dose or exposure merely compounds the problems.[27] Similar questions have begun to trouble once settled conclusions about American occupation troops. Research hinted that participants in nuclear tests during the 1950s might have suffered adverse health effects which appeared only two decades later. In 1978 this possibility prompted the Department of Defense to create the Nuclear Test Personnel Review under the Defense Nuclear Agency. That program was extended to cover Hiroshima and Nagasaki occupation troops in 1979. Subsequent study supports the 1945 view that residual radioactivity posed no threat to occupation forces in Japan.[28] Nevertheless, all doubts have not been allayed, and Defense Department findings have been sharply criticized.[29]

Only the most foresighted observers might have perceived such problems in 1945. Documenting the immediate effects seemed far more urgent, and it was no small task. The Japanese special committee continued working even after the Americans arrived. Its data, in fact, underpinned much of the American effort. Recognizing this fact, Oughterson and Warren sought formal status for the Japanese studies. MacArthur's headquarters approved formation of the Joint Commission for the Investigation of the Effects of the Atomic Bomb in Japan. Initially, the commission comprised Oughterson's army team, Warren's Manhattan Project group, and a Japanese government group under Masao Tsuzuki.[30] A surgeon then teaching at Tokyo Imperial University, Tsuzuki led one of the early surveys in Hiroshima. Ironically, he had studied in the United States before the war and had published on the acute effects of ionizing radiation.[31] A member of the Manhattan Project Atomic Bomb Investigating Group met him in Tokyo and saw a copy of his twenty-year-old study. "When I handed the thesis back to him, he slapped me on the knee and said, 'Ah, but the Americans—they are wonderful. It has remained for them to conduct the human experiment.' "[32]

The Japanese-American joint commission began its work in Nagasaki just before the end of September. A typhoon delayed the start of the Hiroshima study until 12 October. Meanwhile, the joint commission acquired a new member when still another American group reached Japan. Eager for firsthand data on the bombings, the navy sent a technical mission which included a medical unit under Shields Warren. A noted Harvard pathologist and cancer expert, Warren served as naval reserve captain during the war. Established in Nagasaki, the navy team became the fourth member of the joint commission. Clinical, laboratory, and autopsy reports, most of them obtained by Japanese doctors in the first month after the bombings, furnished the bulk of the data. The commission's major effort centered on finding the records, making certain they were correct, and putting them into standard, translated format.[33]

Record making, however, was not the sole work of the American scientists. Physical examinations of still hospitalized patients expanded and updated the collected records. Outpatients at clinics in Hiroshima and Nagasaki added more data. So did a large sampling study which included apparently healthy survivors. Radiation surveys of both cities during October and November by the naval technical mission confirmed and extended the Manhattan Project findings.[34] Possible effects of ingested radionuclides prompted another study. A member of Oughterson's team, Samuel Berg, had worked with Harrison Martland on the New Jersey radium dial-painter cases. He wondered whether fall-out near Nishiyama reservoir had harmed Japanese who consumed contaminated food or water, but nothing he found "could be definitely attributed to the radioactive fall-out from the Nagasaki bombing up to the time of the survey. Follow-up studies on some of the Nishiyama group, and on animals, failed to change this conclusion."[35] Recent studies have attributed to the 1945 attack elevated levels of at least one fission product (cesium 137) in Nishiyama residents, but they found no resultant abnormalities either clinically or in laboratory tests.[36]

Its studies completed, the joint commission dissolved in late December. Having gathered what data they could, the last of the American groups left for home. Specimens and records went to the Armed Forces Institute of Pathology in Washington for further study. Oughterson and his colleagues completed a secret report by September 1946. Five years later, the Atomic Energy Commission released a public version, six volumes of reports and analyses. Oughterson and Shields Warren published a one-volume summary of this larger work in 1956.[37] The Japanese Special Committee for the Investigation of A-Bomb Damages meanwhile sought to publish the data collected immediately after the bombings. Limited funds and other afflictions of a war-ravaged society imposed delays. Occupation policy, however, erected the biggest roadblock. Until the signing of a formal peace treaty in 1951, MacArthur's headquarters sharply restricted Japanese research and publication in any area related to nuclear matters. A summary of the Japanese data appeared in August 1951; the full two-volume report followed in May 1953.[38]

Professional and public knowledge of the acute effects of nuclear attack long preceded the official reports.[39] Journalistic accounts appeared quickly.[40] Medical reports followed, the first finding their way into print soon after study teams returned from Japan.[41] The huge mass of data, much of it never published, conformed to the largely expected pattern of damage inflicted by massive single doses of gamma or X rays: skin burns, hair loss, blood damage, nausea, death.[42] Presumably, however, those who survived would also display later effects, perhaps graded according to degree of exposure. Such long-term effects were much less well known and therefore far more interesting; genetic effects were perhaps the most interesting of all.[43] Much though they might deplore the circumstances, Japanese and American scientists alike perceived the disaster to present, as Secretary of the Navy James Forrestal explained to the president, "a unique opportunity for the study of the medical and biological effects of radiation."[44] Careful study of long-term effects in a large population exposed over a broad spectrum of dose could add greatly to knowledge of human effects. From such concerns grew the Atomic Bomb Casualty Commission.[45]

TOWARD CROSSROADS

Trinity and the bombing of Japan had displayed an awesome weapon. Militarily, however, questions outnumbered answers. America's postwar armed forces found themselves with a weapon seeming to challenge all past military thought. Yet planners knew distressingly little about the strictly military effects of atomic bombs. Strategic bombing, a doctrine in search of a weapon since the 1920s, might now have what it needed. The Commanding General of the Army Air Forces asked the Joint Chiefs of Staff in September 1945 to use Japanese ships for an atomic-bomb test. Memories of Billy Mitchell's victorious 1921 war against the navy—bombing and sinking an aged German battleship off Virginia—no doubt influenced the navy's response.[46] The navy, too, admitted a need for tests: "The appearance of the atomic bomb has made it imperative that a program of full-scale testing be undertaken to determine the effects of this type of bomb, both underwater and above water, against ships of various types."[47]

In October the Chief of Naval Operations suggested a series of tests, overseen by the joint chiefs rather than the air force. He also proposed adding surplus American vessels to the target fleet, and he insisted on navy and army participation in planning and testing.[48] The joint chiefs planning staff studied the key questions—what needed testing, who should be in charge—and furnished some answers in December. The plan called for three tests: first an air burst, then a bomb exploded at or just beneath the water's surface, last a deep underwater detonation. The first three letters of the alphabet provided the names for the tests: Able, Baker, and Charlie. To squelch any question about fairness, a joint task force should conduct the tests. Operating under the joint chiefs, it would draw its staff from all the services as well as from civilian laboratories. An equally broad-based evaluation board, appointed by the joint chiefs, would assess the results. The joint chiefs approved the plan and submitted it to the president.[49]

Military planners sought to learn more about the broad problems atomic bombs posed. How, for example, must the structure and function of armed forces change in response to the new weapons? What effects might such changes require in strategy and tactics? In more concrete terms, they hoped to acquire data on the technical aspects of nuclear war. Primary concern centered on naval design and construction. Thus the target fleet comprised a varied sample of modern naval vessels and merchant ships, arranged to provide graded effects from the burst. The tests also offered a chance to see what happened to aircraft, trucks, and other military equipment disposed about the ships. Munitions, fuel, and normal gear remained in place. Instruments throughout the target fleet furnished physical data. The ships, of course, would remain unmanned during the test, but thousands of rats and hundreds of pigs, goats, mice, and guinea pigs served as stand-ins for study of biological effects.[50] Basically, however, science took second place. "Tests should be so arranged as to . . . acquire scientific data of general value if this is practicable," directed the joint chiefs. Valuable though those data might be, gathering them was not to jeopardize more direct military goals: "the determination of the effects of atomic explosives against naval vessels in order to appraise the strategic implications of atomic bombs including the results on naval design and tactics."[51]

When President Truman approved the plan, the joint chiefs named Vice Admiral W. H. P. Blandy to command the task force. Former chief of the navy's Bureau of Ordnance and one-time commander of amphibious forces in the Pacific, Blandy had been an early choice for the post. He returned to Washington in November 1945 as deputy chief of naval operations for special weapons to help plan the tests.[52] He proposed calling the operation "Crossroads": "It was apparent that warfare, perhaps civilization itself, had been brought to a turning point by this revolutionary weapon."[53] Once named task force commander, Blandy moved quickly. Systematic planning and the search for a test site had not waited for the president's approval. On 21 January 1946, ten days after he assumed command, Blandy presented the joint chiefs with his plan of operations. He listed target vessels and their arrangement, outlined the structure of Joint Task Force One, discussed means of dealing with security, safety, and other matters, and named the time and place: mid-May 1946, Bikini atoll in the Marshall Islands.[54]

Radiological safety strongly affected the choice of both date and site. Good weather, predictable winds and ocean currents, distance from population centers—all helped minimize chances that test workers, observers, or hapless bystanders might suffer undue exposure.[55] The natives of Bikini numbered only 162, too few to present relocation problems; they agreed to move and landed on Rongerik atoll in March. Even as navy ships carried them to their new homes, Joint Task Force One discussed evacuation plans for Rongerik and other nearby atolls should shifting winds threaten them with test fallout.[56]

Bikini was a typical mid-ocean coral atoll, a reef ringing a lagoon of well over 200 square miles, nowhere deeper than 200 feet. It offered ample protected anchorage for both target fleet and support ships. Irregularly spaced along the reef rose twenty-six low-lying islands. The few square miles of dry land were enough for camera towers and recreation, but task force members lived and worked chiefly aboard ship. Nearby Kwajalein and Eniwetok atolls provided required bases for air support. Bikini had only two real drawbacks for Crossroads: its distance from the mainland—4,500 air miles from San Francisco, 2,500 from Honolulu—made extraordinary logistic demands; and its highly humid climate played havoc with delicate electronic and photographic gear. On balance, however, Bikini matched Crossroads' needs far more closely than any other site surveyed.[57]

Blandy tapped William Parsons, newly promoted rear admiral and former Los Alamos Ordnance Division leader, as his deputy for technical direction. Parsons formed his own technical staff and also chose two administrators. To direct test measurements Parsons named Ralph A. Sawyer, graduate dean at the University of Michigan and former laboratory director at the Dahlgren Naval Proving Ground in Virginia.

Los Alamos formed a new technical division, B-Division, to prepare, fire, and measure the results of the bombs.[58] Radiological safety also fell under Sawyer's purview. The Manhattan District took care to distinguish measurements required by safety from other Los Alamos work, defining

> "radiological" measurements as those dealing with hazards to health and "radiation" measurements as all other measurements of radioactivity. . . . Radiological measurements include monitoring of aircraft and ships, radiological survey of lagoon to determine diffusion characteristics, monitoring radioactive cloud and radioactive water after test, and other measurements or technical procedures having direct bearing on determination of hazards.[59]

Radiological safety had been among the first topics the joint chiefs planners discussed. By early December, their meetings with the Manhattan Project high command completed the basic outline of health and safety functions for the projected tests. Blandy and Groves themselves met in January. Responsibility for radiological safety, they agreed, belonged to the Manhattan District. Groves's chief medical advisor, Stafford Warren, headed the radiological safety section under Sawyer. At the same time, Warren served as Blandy's radiological safety advisor.[60] Personnel safety "received primary consideration throughout," Blandy later remarked in summing up Operation Crossroads. "The most strenuous and meticulous efforts were directed toward protection, not only from . . . effects immediately accompanying the detonation, but also from the ensuing and continuing radioactivity."[61]

Warren found Colonel James P. Cooney, radiologist and Regular Army doctor since 1927, to direct instrument repair, dosimetry, and analyses. Los Alamos Health Group provided Louis Hempelmann, James Nolan, and other experts to oversee various phases of fieldwork. From Rochester via Los Alamos came physical chemist Herbert Scoville, Jr., to handle data gathering and plotting for the section. The University of Rochester also sent the man who headed the wartime uranium toxicology project, Andrew H. Dowdy, to chair Stafford Warren's Medico-Legal Advisory Board.[62] Meeting from late June to mid-August on no fixed schedule, the board varied in membership and purpose.

> Initially it served to reassure Col. Warren that the safety measures adopted by RadSafe were such as to attract no justifiable criticism, and to give what assurance was possible that no successful suits could

be brought on account of the radiological hazards of Operation Cross-roads.

Later the board defined for itself a larger role, especially after Test Baker raised a host of unanticipated problems.[63]

Rear Admiral T. A. Solberg came from the navy's Bureau of Ships to head Parsons's second technical division. As director of ship material he oversaw a 10,000-man effort to assess and repair test damage. Included in that number was a second safety team. The navy's Bureau of Medicine and Surgery seconded George M. Lyon to head the so-called medical group. A 1920 graduate of Johns Hopkins Medical School, Lyon practiced pediatrics until World War II brought him a reserve captain's commission in the Navy Medical Corps and a chemical warfare assignment. His Crossroads medical group actually had two largely distinct functions: dealing with all the nonradiological hazards likely to confront the task force and running a biological research program. Like Warren, Lyon wore a second hat, serving as the task force safety advisor. He and Warren worked closely together, Lyon in Washington organizing the early safety program. Among the first tasks in which he and Warren joined was writing the safety plan for Joint Task Force One.[64]

THE CROSSROADS SAFETY PLAN

The Crossroads "Safety Plan" appeared as Annex E to Blandy's operation plan, one of twenty-nine annexes running in all to several thousand closely printed pages. Annex E included ten appendixes and totaled nearly a hundred pages; it covered precautions against more mundane hazards like fire as well as radiation. Radiological safety, however, rad-safe in common usage, was the major concern. The safety plan defined the basic mission of the radiological safety section: "To protect personnel from the hazards peculiar to the use of the atomic bomb during Operation CROSSROADS and to enable personnel to return safely to the target area at the earliest possible moment."[65]

Protection against direct bomb effects—blast, heat, light, prompt radiation—presented no serious problems. As at Trinity, distance would ensure safety; the nearest task force units were no closer than 15 miles at the moment of detonation. Reconciling safety with quick return to the target area posed the real challenge. On the one hand,

There will be a large staff of technical and scientific observers who will be eager to board the target ships as early as practicable and safe. Further, it is of vital importance to the operation that the salvage parties be able to approach Target Ships as soon as possible in order to prevent secondary damage from completely masking the results of the test.

On the other hand, "the detonation of an atomic bomb will present many hazards, some of which are not well understood."[66]

Radioactivity induced by neutrons, airborne fission products, and fallout were among the hazards foreseen.[67] From the outset, however, the subsurface test called Baker presented the most unknowns and raised the deepest concerns. One early analysis warned that

the water near a recent surface explosion will be a witch's brew, and this will be true to a lesser extent for the other tests. There will probably be enough plutonium near the surface to poison the combined armed forces of the United States at their highest wartime strength. The fission products will be worse.[68]

Blandy's first Crossroads information circular in January 1946, while avoiding language so colorful, judged "it will undoubtedly be some weeks before the lagoon and target ships are again habitable."[69] Further study changed little. The April draft "Re-Entry Plan for Test Baker" expected "radioactivity . . . much greater and more persistent than after Test 'Able.' "[70] A May report predicted "safety problems . . . aggravated by this greater amount and higher concentration of radioactive material and by the consequent larger persistence in the water of the lagoon, and on the central target ships."[71]

Against all hazards the safety plan stressed two key defenses; "*Detection* and *avoidance* provide the best protection. This is the basis of the Safety Plan as far as radiological hazards are concerned."[72] Detection relied on monitors and their instruments, Victoreen model 263 Geiger counters and model 247 ionization chamber meters plus a number of experimental devices supplied by the Manhattan District. Systematic reconnaissance begun shortly after firing would include aerial surveys in navy flying boats over the lagoon, ocean patrols by navy destroyers for fallout both upwind and downwind from the blast, sampling by drone aircraft and boats, lagoon patrols in small boats and landing craft, and cloud-tracking by air force B-29s. Monitors would also join the initial boarding teams clearing target ships for salvage work. All teams would be linked by radio to Warren and the rad-safe control unit aboard the task force flagship, USS *Mt. McKinley*.[73] Analyzed and mapped, field data would allow the unit to discharge its "ultimate, complete and vitally important responsibility" to advise the task force commander "as to the location, severity and probable significance of hazardous areas, and advising him on action for the safety of personnel."[74]

Safety planners foresaw little or no hazard to most task force members, whose duties kept them well out of harm's way. Individual safety measures applied only to the small fraction judged most likely to be exposed, such as radiological patrols, salvage units, and air crews. Monitors carried film badges and pocket dosimeters. Badges were also issued to some or all members of the crews they worked with. In general, badges were issued daily and processed immediately upon return, the special task of the photographic dosimetry (photometry) unit. Like other rad-safe technical service units, photometry was housed aboard the USS *Haven*, a navy hospital ship assigned and equipped for this purpose.[75] The Crossroads film badge used a standard commercial X-ray emulsion. Unlike wartime badges, it had only a single film with a range of 0.04 to 2.0 roentgens. Black paper covered the film, crossed lead strips served as gamma-detection filters, and a special tropical envelope protected each film packet against Bikini's humid air. Some 10,000 film badges were used in Crossroads, 4,000 in the first test and 6,000 in the second. The Manhattan District supplied them all, as well as a small number of casualty badges, special multifilm badges to cover higher exposures on ships.[76]

Exposure standards for Crossroads conformed to Manhattan District usage: the basic standard was the same 0.1 roentgen per day adopted by the Chicago Health Division in 1942 and later applied throughout the project. The statement in the Crossroads safety plan reflected the still not fully resolved issue of permissible exposure versus tolerance; it also assumed that the daily limit might be indefinitely prolonged without harm: "The maximum allowable dose or tolerance for daily exposures over a long period of time is 0.1 roentgen."[77] Such a long-term standard, however, would not meet the immediate needs of testing. Just as the Los Alamos Health Group proposed special limits for Trinity, so Crossroads planners pondered a short-term standard for the Bikini tests: "It is considered that radiation exposure of 15 R per 24 hours is dangerous. Radiation exposure of 50 R per 24 hours is definitely not to be encountered."[78] Eventually, the radiological safety section settled on somewhat more restrictive standards:

> For purposes of safety in this operation, it is considered that an individual should not have a total exposure of over 50 to 60 roentgen in two weeks. If an individual receives 10 roentgen in one day, or 60 roentgen in two weeks he will be withdrawn from active participation in the operation.[79]

This higher limit would require direct clearance from rad-safe control upon request from the field: "Special situations may permit the assuming of a calculated risk in order to let certain key personnel enter a hazardous area to make highly desirable observations when the total amount of radiation is less than 10 roentgen units."[80]

Initial plans called for a sixty-man radiological safety section, half from the Manhattan District, half from the armed forces. This latter group—mostly Army Medical Corps officers, but also navy, air force, and Public Health Service doctors, and a chemical warfare officer—required special training. In mid-January 1946 they began an intensive eleven-week course: at Oak Ridge on basic physics and plant safety routines; at Chicago and Rochester on biological effects; at Los Alamos for field training; at Berkeley on exposure from ingested radionuclides.[81] Working on safety plans meanwhile showed Warren and his staff how badly they had misjudged the demands Operation Crossroads would impose.[82]

Radiological safety required many more monitors and other experts than first believed. Even draining the Manhattan District of almost every trained person left the section shorthanded. Warren felt compelled to call upon former workers who had already returned to civilian posts. Many responded, but few eagerly. They demanded and received a promise of strict time limits on their service abroad.[83] "This is a bottleneck, the severity of which can hardly be estimated accurately at this time but it does impose a *real* and *serious* threat to the essential safety requirements of the operation," warned the radiological safety section. "If personnel *adequately qualified for radiological safety reconnaissance are not available . . . a serious delay may be occasioned, or the plans of the operation may have to be changed materially.*"[84] Critical in February, the problem became much more serious late in March when Crossroads was postponed.

CROSSROADS DELAYED

Crossroads planners never enjoyed the freedom that secrecy lent Trinity and the attack on Japan. As early as 10 December 1945, President Truman announced the nation's intent to conduct tests in the near future. Immediately after the joint chiefs approved his plans, Blandy likewise went public. That was late in January, just weeks after the United Nations received American proposals for international control of nuclear weapons.[85] Policy in this broad sense had played little part in Crossroads planning or timing. Appearances, however, suggested otherwise, both within the United States and abroad. Reacting to charges that tests would merely prove what their military sponsors wanted to prove, the president formed his own evaluation commission, distinct from the joint chiefs evaluation board.[86] Diplomacy also affected Truman's sudden decision in March to postpone the first test from 15 May to 1 July. Practical politics entered as well: he "didn't want a lot of Democratic congressmen out witnessing the test when their votes were needed here in Washington."[87]

Postponement eased some Crossroads problems but exacerbated others. Blandy had good reason for choosing 15 May, despite strains early testing might impose. Joint Task Force One needed tens of thousands of men at a time of rapidly dwindling force levels. Each day's delay compounded the problem, the more so as untrained recruits too often replaced seasoned crewmen. As early as the beginning of February the Chief of Naval Operations learned that 80 percent of crew members in ships assigned to Joint Task Force One had less than six months to serve, with few replacements available.[88] Postponement forced the navy to begin stripping the active fleet to man the task force, but skilled manpower remained in short supply.[89] Worse, many of Blandy's civilian scientists expected to return to campus for the fall term; shifting the tests to midsummer threatened their participation.[90]

For the radiological safety section, that possibility quickly became more than threat. A worried Warren telephoned Washington early in April. He had so far managed to sign but three-fifths of the 125 civilian monitors needed for Able, only one-fifth of the 125 Baker required. "Due to uncertainty of dates of tests, and possibility of later delays from the White House, they are increasingly reluctant to sign up; or indeed actually withdraw from contract." Military officers and enlisted men for the rad-safe team seemed less a problem, but they represented the unit's second-line strength and its support. All the "first grade monitors," were civilians. Warren found few willing to remain in the Pacific past August.[91]

"Postponement of the tests forced many civilians to cancel," Warren later reported, which provoked "a further frantic search for replacements."[92] He recruited still other civilians and obtained seventy more officers, this time mostly from the naval reserve.[93] In the end, he mustered 381 men for Test Able, "just enough to meet the requirements."[94] Organization and training, however, suffered. The bulk of Warren's section reached Bikini in mid-June, well behind the rest of the task force and only two weeks before the scheduled firing date. Equipment shortages now compounded his problems. The six-week delay also meant that many who came for Able would not stay for Baker. Replacements fell well short of losses, and training new team members imposed further burdens.[95]

Improvisation and hard work sufficed, if barely, to handle these problems. Another delay-imposed challenge raised more serious doubts. Although the northern Marshalls enjoy fair weather almost year-round, summer brings the equatorial front. During late June and more often in July, winds from north and south of the equator converge over the islands. Occasional clear days occur, but heavy clouds, fickle winds, and rainsqualls become the rule.[96] The prospect of exploding a fission bomb at the wrong time presented "an all too possible nightmare," in the words of one of Blandy's chief weather experts, naval Captain A. A. Cumberledge. Detonation would produce "the equivalent of tons of radium floating loose in the atmosphere in deadly concentrations. To guarantee that the tests would not be suicidal," Cumberledge added, the task force must "make sure that the winds [would] carry the contaminated atmosphere away from personnel participating in the tests."[97] Blandy himself phrased it more tersely: "Our weather forecast . . . was not just a matter of 'fair and warmer'; it was a matter of life and death."[98]

When Joint Task Force One reached the Marshalls in early spring 1946, it planted army and navy weather stations throughout the region. Limited land meant using shipboard stations as well. Reconnaissance aircraft, however, played the key role. Specially equipped to gather weather data, air force B-29s and navy seaplanes flew twice daily on scheduled missions lasting up to fourteen hours. From Hawaii, fleet weather central sent data from stations across the North Pacific to augment local readings. Crossroads had its own weather central on Kwajalein to collate, map, and analyze the data. Seconded by Cumberledge, air force Colonel Benjamin J. Holzman headed the aerology section under the assistant chief of staff for operations. Aboard the *Mt. McKinley*, the staff aerological unit used reports from Kwajalein to prepare daily forecasts. Each morning Blandy had a formal forecast for the next thirty-six hours, the time he needed to decide whether or not to fire. Installing and checking instruments, clearing all workers from the target, and moving ships and aircraft to test stations required twenty-four hours. Radiological surveys after the test needed a full day of clear weather.[99]

Aerological morning reports stressed detailed forecasts of cloud cover and winds. These were crucial for Test Able. A B-29 flying at 30,000 feet would drop the bomb; test guidelines required the bombardier to see his aiming point. Clouds predicted to cover more than three-tenths of the sky below the bomber became reason to postpone the test. Accordingly, the forecast included the ratio of sky blocked by clouds at low, middle, and high levels, as well as the heights of each cloud layer. Winds varying widely in speed and direction at different altitudes— all too likely over the Marshalls at that time of year—might deflect the path of a falling bomb. Daily forecasts included wind speed and direction for each 5,000 feet from the surface up to 60,000 feet. For all tests, wind forecasts were vital to fleet safety, since a sudden shift in wind might shower crews with fallout. Confidence rose, however, as routine daily forecasts less and less often missed the mark. Approaching 1 July, Holzman and Cumberledge grew ever more certain that Crossroads could rely on their forecasts. Blandy agreed.[100]

Yet all failed to go smoothly when Joint Task Force One attempted its Queen day operation, a full-scale dress rehearsal as final preparation for Able. On Saturday morning, 22 June, the forecast promised weather good enough for testing next day. Accordingly, Blandy fixed 23 June as Queen day and ordered the exercise to begin. Saturday's effort proceeded on schedule. Late that afternoon, the lagoon now emptied of all save the target fleet and a few planned last-minute ships, the Crossroads weather experts met for their daily review of the morning's forecast. Later data compelled them to alter their prediction. Excessive cloud cover would leave the B-29 too little chance to complete its bombing run next morning. Informed of the bad news at his regular 10 P.M.

weather briefing, Blandy decided to postpone Queen day, but he kept most ships on stations.[101]

Prospects brightened Sunday morning, and Blandy rescheduled Queen day for 24 June. The last ship cleared the lagoon entrance at four o'clock Sunday afternoon. This time the late-afternoon weather conference brought no surprises. Although Monday would be cloudy, the B-29 should have a clear shot at its target—the battleship USS *Nevada* anchored in the center of the target array and brightly painted to enhance its visibility. Following his ten o'clock weather briefing, Blandy confirmed Queen day orders. The planned time of detonation—How hour—would be 8:30 next morning. When the moment arrived, however, clouds obscured the target. Thirty thousand feet above the lagoon, the B-29 named *Dave's Dream* swung around for another try. Three-quarters of an hour later, the second bomb run succeeded. The practice high-explosive bomb burst slightly aft and starboard of the *Nevada*. Blandy flashed the message to the task force: Mike hour, the actual time of detonation, was 9:14. That signaled the vast recovery operation to begin.[102]

Above the lagoon, army and navy aircraft photographed, measured, sampled, and surveyed along assigned courses. Radiological patrol ships began steaming toward their missions within minutes of the blast. The first drone vessels entered the lagoon just before noon to begin plotting the supposed pattern of radioactivity. Other ships followed in waves. In early afternoon the first crews reboarded target vessels. By six o'clock almost the entire fleet once again rode at anchor within the lagoon. The Queen day exercise concluded shortly after noon the next day with the boarding of the last target ships. Inaccurately predicted clouds aside, the performance was very nearly flawless. Joint Task Force One showed itself ready for Able. As 1 July approached, the only real question remained the weather.[103]

6

CROSSROADS

CROSSROADS ABLE

To David Bradley, an army doctor in the radiological safety section, Saturday, 29 June, looked "perfect. After three weeks of bad weather we have hit a fair streak."[1] He, of course, was no expert. Professional optimism was more guarded. The latest weather map showed a vast low-pressure system west of the Marshalls; overcast skies and showers spanned the mid-Pacific from the Philippines to the Carolines. That system, however, appeared unlikely to deepen or move much during the next two days. An upper-air low-pressure trough passed the Marshalls on Saturday. It caused severe thunderstorms, but only a day later and hundreds of miles east of Bikini. The ever shifting equatorial front now ran well to the south, with little chance of infuencing weather in the northern Marshalls through Monday. All things considered, 1 July looked like a fair bet for Able.[2] Still, a degree of uncertainty persisted:

> It has become increasingly apparent that we cannot expect to have on Able Day, the ideal situation wherein the winds are from easterly quadrants at all altitudes. Neither can we with the period at our disposal afford to wait for time to bring us conditions even approaching this ideal. . . . It is accordingly necessary to be prepared for many and varying eventualities.

Joint Task Force One received contingency orders from its commander, Admiral Blandy: if worst came to worst, a brief message broadcast from his flagship, the *Mt. McKinley*, could set the fleet promptly into motion.[3]

The official weather forecast Sunday morning informed Blandy of the next day's favorable prospect:

> Two- to three-tenths cumulus clouds with bases at 1,500 feet, tops at 5,000 feet. No middle clouds. About six-tenths of high cirrus clouds at altitudes above 30,000 feet. Total cloud cover below bombing altitude at target time two- to three-tenths. Winds aloft expected to be

easterly 10 to 15 knots up to 15,000 feet, variable at 2 to 8 knots between 15–25,000 feet and northwesterly 25 to 35 knots above 25,000
feet.[4]

At nine o'clock Sunday morning, 30 June 1946, Blandy signaled Joint
Task Force One his decision. Able day would be 1 July, How hour 8:30
in the morning. Messages alerted higher commands in the Pacific, re-
minding them to allow no air traffic within 500 miles of Bikini after
8:30 Sunday night. Other messages informed Washington of the
planned firing.[5]

Everything stayed on schedule Sunday under bright blue skies dotted
with cottony white clouds unmarred by any smudge of rain. Evening
brought "a marvelous sunset" over Kwajalein, Bradley observed.
"Clouds had formed, as they often do here, directly over the islets of
this atoll, stringing out in a similar broken chain overhead."[6] Cool night
air normally caused increased cloudiness. Such the weathermen expected. They predicted a maximum near dawn over Bikini, but swift
morning clearing. Reconnaissance flights on Sunday, however, found
higher than expected moisture in the air below 5,000 feet. The experts
believed this condition would merely increase predawn clouds, with
no effect on How hour. The ten o'clock weather briefing left that morning's forecast unchanged. Blandy's orders stood.[7]

Reconnaissance planes crisscrossed the sky over Bikini all night,
watching the clouds and testing the air. Radioed reports went direct
to the *Mt. McKinley*, still anchored in Bikini lagoon. The news was
gloomy. Cumulus clouds covered up to eight-tenths of the sky and
towered as high as 13,000 feet. Frequent lightning played about the
lagoon and showers pelted its surface. Weather so extreme diverged
sharply from expectations. Aboard the *Mt. McKinley*, the weathermen
studied the new data and reviewed their forecast. At five o'clock Monday morning they presented a special weather briefing. Despite appearances, they stood by their forecast of decreasing cloud cover after
dawn and no more than three-tenths at shot time. Even as the briefing
proceeded, late aircraft reports reaching the flagship tipped the balance. The predicted trend could already be observed. Blandy saw no
need to cancel the test. Cautiously, however, he slipped How hour to
nine o'clock and broadcast the revised orders at 5:42. Already under
way, the *Mt. McKinley* cleared the lagoon at 6:30, the last ship to leave.
Only the target fleet remained.[8]

Meanwhile, at Kwajalein *Dave's Dream* had waited on the runway for
Blandy's order. No one wanted to chance a return landing with a bomb

aboard if the test had to be postponed. Within minutes of the order, the B-29 lumbered into the air, heading for Bikini. Many other army and navy aircraft with Able day missions were already airborne. The brief delay caused no problems; the planes simply flew pretest stations for the extra half hour.[9] Aboard a navy seaplane that had departed just before dawn, David Bradley checked his instruments. His was one of two planes assigned to radiological reconnaissance. After the shot, it would sweep the air low over Bikini to locate and map the pattern of radioactivity. Each crew member was given dark goggles to protect his eyes from the flash at detonation. Bradley also handed out gas masks and film badges.[10] Conditions looked none too promising at 7:30 when the plane reached its station. "At 3,000 feet we were running into a steady sea of fleecy clouds," Bradley noted. "There was some doubt in our minds whether the drop ship could make out the target."[11]

Dave's Dream arrived over the target array just after eight. Visibility, despite earlier qualms, exactly matched the forecast as the B-29 completed its scheduled practice run at 8:30: clouds covered two- to three-tenths of the sky. The bombardier had no trouble sighting the target.[12] "Bombing conditions were okay," he radioed after the practice run. Bradley thought he detected "a trace of satisfaction in the 'okay.' "[13] Conditions were even better half an hour later, with cloud cover down to two-tenths. Flying a steady 300 miles per hour at 30,000 feet, the bomber released its payload fourteen seconds before nine o'clock.[14] Bradley listened to the countdown from the B-29 he could not see. Finally he heard, "coming up on actual bomb release. Stand by. . . . Bomb away. Bomb away. Bomb away. Bomb away."[15] Following the pattern first devised for the attack on Japan, *Dave's Dream* circled into a shallow dive to get away more swiftly.[16]

Mt. McKinley flashed the message: Mike hour, the actual time of detonation, was 0900.[17] But for Bradley and his crewmates on station 20 miles away, "the bomb went off unannounced." Intervening clouds blocked the flash; they heard no sound, felt no shock. Even when they removed their goggles, they at first saw nothing.

> Then, suddenly we saw it—a huge column of clouds, dense, white, boiling up through the strato-cumulus, looking much like any other thunderhead but climbing as no storm cloud ever could. The evil mushrooming head soon began to blossom out. It climbed rapidly to 30,000 or 40,000 feet, growing tawny pink from oxides of nitrogen, and seemed to be reaching out in an expanding umbrella over-

head. . . . For minutes the cloud stood solid and impressive, like some gigantic monument, over Bikini. Then finally the shearing of the winds at different altitudes began to tear it up into a weird zigzag pattern. Winds high up were from the west and so the head tended to move out over us and menace the live fleet, while shreds of the torn column beneath could be seen moving slowly westward.[18]

Beneath the circling seaplane steamed Blandy's ships, among them the USS *Appalachian*. It carried more than a hundred newsmen. Unlike Trinity, Crossroads was a public event. Broadcast live from the *Appalachian*, however, it found a less than enthralled audience. According to a Philadelphia *Record* phone survey, Crossroads scarcely outdrew a competing baseball doubleheader.[19] Observers on the scene, although undistracted, saw less than they had expected to see. "I think I feel a bit disappointed and let down," one noted in his diary.

The show was not as spectacular or dramatic as we had been led to expect. We were unnecessarily far away and it was impossible to see the target array even with good Navy glasses. . . . The damage was less than a few well-placed aerial bombs would do, that is, as far as we can see now.[20]

Many of the quarter-million words of copy filed from the *Appalachian* in the twenty-four hours after Mike hour sounded a note of relief. Reporters, like the public at large, seemed to William Laurence—the only reporter who had witnessed Trinity and Nagasaki firsthand—"only too eager to conclude that the atomic bomb was, after all, just another weapon."[21]

ABLE RECOVERY

As soon as the *Mt. McKinley* announced Mike hour, aircraft converged on Bikini. The first radio-controlled drone planes plunged into the cloud within minutes, each equipped with special filters and bags to collect particulate and air samples. Manned aircraft approached with more caution. The so-called radiation exclusion sector, or radex, the region off limits to task force members for safety reasons, fell downwind. Initially, that meant almost anywhere west of the target; variable winds at different altitudes did not allow rad-safe control aboard the *Mt. McKinley* to define a narrower sector.[22] Approaching Bikini, aircrews became the first concern of the rad-safe section and its leader, Stafford Warren. Monitors like Bradley rode in each plane, and all crew members wore film badges. Many pilots had themselves practiced flying over Trinity site in New Mexico to learn something about airborne monitoring.[23] "All of these pilots," Warren believed, "had enough understanding of the problem to know when to take evasive action."[24]

Surface winds blowing mostly westward clearly removed any danger of fallout over Rongerik, new home of the transplanted Bikinians, more than 100 miles east of their former home. After riding out the test in navy boats ready for quick evacuation, they could return to shore. So ordered Blandy just after ten o'clock. At Eniwetok, 200 miles west of Bikini, however, five C-54 cargo planes assigned as an evacuation unit remained on standby until early afternoon. Destroyer patrols, which began fallout surveys both upwind and down from Bikini shortly after the shot, found nothing significant on the surface all day.[25] The Able cloud, in fact, dissipated far more swiftly than had been expected; variable winds above 15,000 feet diffused it in all directions. Four B-29 crews had the mission of tracking and photographing the cloud as long as they could see it—that is, until darkness fell or the cloud vanished. That mission ended at two in the afternoon, when nothing more of the cloud could be seen. Blandy could then call off the alert at Eniwetok and also lift the ban on nontest flights within 500 miles of Bikini. Instrument-carrying aircraft located remnants of the cloud that night, well north of Eniwetok.[26]

Such distant concerns played small part in Able day activities. Bikini itself remained the focus of most eyes. The two navy radiation reconnaissance seaplanes had the first close-up look. Like the more distant watchers at sea level, they saw less than they had expected. Against the sometimes wild predictions everyone had heard, reality paled.[27] Despite fires on many ships, damage seemed "surprisingly light," Bradley commented. "Expecting much more dire and dramatic events our crew was disappointed."[28] Yet compared with any past naval experience, the havoc wreaked by that single bomb, even allowing for the special target placement of ships, was substantial.[29]

The first damage report to the joint chiefs from Bikini early that afternoon offered a compressed and matter-of-fact picture:

> Transports GILLIAM and CARLISLE sunk. Destroyer LAMSON capsized. Jap cruiser SAKAWA, light carrier INDEPENDENCE, cruiser PENSACOLA, and submarine SKATE considerable damage. Jap battleship NAGATO, battleship NEVADA, transport CRITTENDON, YO-

160 and LCM-1 slight damage. Other ships no visible damage reported from air.[30]

Although the target fleet hardly emerged unscathed, the much remarked slightness of damage was no mere figment. The fleet did, in fact, suffer less than expected, and the still floating *Nevada* offered a clue. As Bradley's seaplane flashed low over the center of the target array, he spotted the battleship, its gaudy paint badly singed only aft. It "turned out to be some distance from the most intense radiation," he noted.[31]

The bomb missed its target by a wide margin. Practice bombs had rarely fallen more than 500 feet off bull's-eye, but the Able day bomb exploded over a point three to four times that far from the *Nevada*. Why? Speculation abounded, the more so because Blandy declined to concede the error for more than a week, pending full study. Study failed to pinpoint the reason. Paul Tibbets, pilot of the Hiroshima bomber whose crew failed to win the Crossroads mission, blamed what he called the "fiasco at Bikini" on the winning crew's incompetence.[32] Others saw "something fishy" in the whole matter.[33] Whatever the reasons (no final answer ever appeared), the misplaced bomb partly explained the less than expected damage. It exploded at the correct height of 520 feet, but over a more open portion of the target array; only one ship was within 1,000 feet. Fewer ships close to the center of the burst meant lesser damage overall.[34]

Assessing damage, however, could wait, as immediate concern centered on the radiation picture in Bikini lagoon. Obtaining the first systematic readings was the prime mission of the radiation reconnaissance seaplanes. They had to follow a strictly patterned flight plan: back and forth crosswind, from well east of target center to an equal distance west, first at 2,000 feet, then at 1,000, and finally at 500. The first plane launched its pattern just before ten o'clock.[35] Bradley, monitor on the second plane, listened closely to the radioed reports. Conditions proved no worse than had been expected, as his own Geiger counters soon showed. Radioactivity seemed most intense where induced in the central target ships themselves—"each one seemed to catch us in a beam, as though from a searchlight, as we passed overhead," he remarked—but smoke from burning ships also brought the Geiger counters to life.[36] Overall, the readings held no surprises.

Recovery efforts in the lagoon were already well under way when the seaplanes completed their survey early Monday afternoon. Like the first cloud samples, the first water samples came from drones. Radio-controlled boats conned from aircraft entered the lagoon less than an hour after the shot, while their tender hovered beyond the reef. A task force destroyer fitted with radiation-monitoring gear swept the lagoon entrance and reported it free of contamination. At 10:50 the first manned boats passed through the channel into the lagoon to begin mapping its waters for radiation. Reading Geiger counters as they went and sending frequent reports to the *Mt. McKinley*, monitors approached the upwind, eastern rim of the target array. From several points around the rim, they then began slowly working their way among the target ships toward the center. By one o'clock rad-safe control fixed the eastern limit of the danger area as a line running through the center of the target array. Radiation patrols began working westward along the northern and southern edges of the danger area to outline its full extent. Night fell, however, before they could close the circle. Radioactive waters mainly surrounded target vessels near the center; as early as 2:30 Blandy could declare the lagoon safe for all ships, although high readings in the planned anchorage forced most of them to use temporary berths overnight.[37]

Salvage units followed the radiation patrols into the lagoon. Fighting shipboard fires in the cleared areas became their first order of business. The ten initial boarding teams boarded the first damaged ships as soon as they received clearances. Two monitors led each team. Clad in heavy coveralls, boots, gloves, and gas masks, they surveyed debris-strewn decks and such areas belowdecks as they could safely reach. They decided whether to declare the vessel "Geiger sweet"—radiologically safe enough to permit salvage work to begin—or "Geiger sour"—unsafe for boarding. By nightfall the teams had cleared 18 target vessels. Only one major fire still burned, aboard the yet "Geiger sour" aircraft carrier *Independence*. When Blandy forwarded the day's final damage report just before midnight, the score had climbed. The toll now stood at 4 ships sunk, 7 badly damaged, 3 moderately, and 24 slightly.[38] Writing to his wife at the same hour, an elated Stafford Warren stressed other results: "Nobody got hurt & nobody got more than tolerance & our organization functioned smoothly tho at times hectically."[39]

Salvage and cleanup resumed early Tuesday morning. Able day plus one produced no surprises. Work proceeded smoothly as rad-safe control cleared the center of the target array for salvage units. Radioactivity in lagoon waters was largely neutron-induced; virtually all fission products and unfissioned plutonium rose with the cloud and scattered widely on the winds. By noon Tuesday no water anywhere in the lagoon read above 0.1 roentgen per 24 hours, the agreed tolerance standard for Joint Task Force One. Radioactivity aboard the target fleet was likewise mostly neutron induced and, with few exceptions, very low. Natural decay quickly removed the hazard, and no target ships required decontamination. Only 13 ships on Tuesday produced readings higher than 0.1 roentgen per 24 hours.[40]

Radiological patrols found the job growing ever more routine. As salvage units completed their work, crews reboarded the target ships to restore them to normal order. Several badly damaged ships required beaching, however, and a fifth ship sank. Radiologically unsafe, the Japanese cruiser *Sakawa* was barred to salvage units and succumbed to progressive flooding. At day's end only ten target vessels remained unmanned, some still too "sour," others with machinery too badly damaged to support crews. Operational duties over, the rad-safe section left the *Mt. McKinley* Tuesday evening for its home base, the *Haven*. The well-equipped navy hospital ship furnished the air-conditioned setting for the section's technical work. Processing film badges became the big job in the days just after Able.[41]

Badges went to only a small fraction of the task force, but the job kept the photometry group working late into the night. Group leader Gerhard Dessauer reported a total of 4,000 daily badges processed for Able; half of them were actually worn by members of the task force, the other half were spotted about the target ships as data collectors. Sensitive to a range from 0.04 to 2.0 roentgens, Able personnel badges showed only ten readings at or above 0.1 roentgen. That was the accepted daily limit adopted as standard for Operation Crossroads. In two instances, readings hit the top of the film's range; both appeared anomalous, since others in the same situation read much lower. The other high readings all fell between 0.1 and 0.2 roentgen.[42] In Test Able, Dessauer concluded, "no personnel of the task force obtained a physiologically significant radiation dose."[43]

Contamination became a modest concern only with respect to the drone aircraft: "Returning from Test ABLE," Technical Director William Parsons observed, they "were radiologically hot at 100 yards."[44] Almost a week after Able, they remained " 'isolated' as they are still highly contaminated with radioactive materials."[45] This problem was only a minor one. Although further cleanup and repair work on target ships lasted for days, Able had passed into history. Only the never explained bombing error and a single drone aircraft lost at sea marred a perfect test. Joint Task Force One swiftly shifted its efforts to Baker. Personnel turnover and other problems scarcely delayed progress toward the scheduled target date, 25 July 1946.[46]

The radiological safety section issued its updated plan for Baker on 15 July.[47] The section also offered to provide badges to those who might board ships after the test "and want to keep track of their total exposure to gamma rays."[48] Preparations climaxed in a dress rehearsal for Baker on William day, 19 July. Although bad weather curtailed air operations, William day met its goals. Weather played not quite so crucial a role in Baker as in Able. Radioactivity from an underwater test would not rise so high, meaning upper-level winds mattered less. Visibility good enough for test photography also imposed fewer constraints than aerial bombing; the Baker day bomb would hang ninety feet below an anchored barge to be detonated by radio signal.[49]

TEST BAKER

Even though it mattered less, weather still dictated the firing decision. The forecast on Tuesday morning, 23 July, promised conditions much like those that produced good weather for Able. That prospect weakened, however, when surging trade winds carried the equatorial front to Bikini on Tuesday night. Deteriorating weather, thunderstorms, and heavy showers throughout the region left the experts doubtful. Still, they could see no certain reasons to alter their promise of clearing weather by Thursday. They repeated it at the Wednesday morning weather meeting but reserved final judgment, as later reports might force them to change the forecast. Blandy issued orders for Baker next morning, if the evening weather conference confirmed the morning forecast. It did. Although Blandy and his staff could see lightning playing to the north and hear rain drumming on the *Mt. McKinley* while they conferred, the orders stood.[50]

Again the experts were right. "The weather is fine," noted an observer on the USS *Panamint*, at sea 12 miles from Bikini Thursday morning.[51] How hour, the planned moment of detonation, was 8:35. In the *Panamint*'s wardroom at breakfast, a certain nervous humor prevailed. Previous bombs had all exploded in air, only Trinity near enough the ground to make fallout a problem. Everyone knew Baker promised something new. This underwater explosion would instantly turn un-counted tons of water and coral into radioactive vapor. At the very least it would throw out huge waves and, as the vapor condensed, drop radioactive rain over the target area. Test planners, of course, believed the hazards modest. But groundless as such fears proved, no one could flatly deny the wilder stories all had heard—the entire fleet destroyed, the islands swept clear by a giant tidal wave, the ocean itself ignited in a chain reaction.[52]

Detonation occurred on time. At 8:35 A.M. How hour became Mike hour, the actual rather than the planned moment. Although both Bikini bombs were identical to the bomb dropped on Nagasaki and both produced the same announced yield of 23 kilotons, Baker matched its billing far more closely than Able.[53] "As a spectacle, Baker made up for more than all our disappointment at the sight we saw at Able," one watcher entered in his diary. "It was tremendous."[54] Bradley once again saw the test from the air, on station 15 miles northeast of the target. The sight beggared description, too sudden and too huge to grasp. The explosion

> seemed to spring from all parts of the target fleet at once. A gigantic flash—then it was gone. And where it had been now stood a white chimney of water reaching up and up. Then a huge hemispheric mushroom of vapor . . . rapidly filled out in all directions.[55]

Many echoed the verdict of the task force historian: "A giant and un-precedented spectacle."[56]

Erupting from a dome of spray, the pillar of water hurtled more than a mile into the air. Nearly half a mile across, the column was hollow, a wall of water 300 feet thick around a core of jetting gases which burst through the top to form the cloud, more cauliflower than mushroom to many eyes. The column began to collapse within seconds, dropping a million tons of water back into the lagoon. From its base surged a dense wall of mist. A doughnut-shaped cloud 1,000 feet thick and 3 miles across blanketed the target fleet. Expanding and rising, it merged with the cauliflower cloud.[57] Seen from the *Panamint*, "the clouds became darker and denser and completely blotted out the whole horizon in the direction of the target array."[58] Thickening clouds began dropping rain which lasted for nearly an hour, returning the last of its water to the lagoon.

Recovery efforts for Baker, like Able, began with drones, but the first hints of trouble came from the two navy reconnaissance seaplanes. As more intense radioactivity was expected from Baker than from Able, they began at 4,000 rather than 2,000 feet.[59] By the time the first plane reached 2,000 feet it was "running into trouble—trouble in the form of radioactivity so intense as to be safe for only a few minutes," Bradley noted. Abandoning the pattern, both planes awaited new orders, which soon came.

> Discontinue . . . surveys. We are thankful for that. It is not that we were in any immediate danger. But with radiation so intense at such an altitude, that at water level would certainly be lethal. And this wasn't just a point source, it was spread out over an area miles square.[60]

Confirmation soon came from the boats probing the lagoon.

Radiological patrol boats moved into the lagoon within two hours after the shot, but Able day tactics worked no better for boats than for aircraft. Baker day surface winds were light, 5 miles an hour toward the northwest, making the cloud slow to disperse. Approaching the target array upwind, patrols found clear waters around a few of the outermost ships on the southern and eastern rim. Seven ships were cleared for boarding; monitors from the initial boarding teams confirmed topsides free of activity, but that proved the day's bright spot. When patrols tried to move toward target center, they quickly ran afoul of intense radioactivity. Notifying rad-safe control to call off the salvage units, they withdrew.[61] Film badges recorded thirty-six mild overexposures among members of patrol crews on Baker day; most fell below 0.2 roentgen, and the highest was 0.25. Contamination, however, did force one crew to leave its boat that evening, and almost the entire lagoon remained off limits when Baker day ended.[62]

Radioactivity so intense came as a surprise. Predictions of the amount of water thrown into the air hit quite close to the mark. Planners also expected most of the bomb's active products to remain trapped in the lagoon. Too low and too moisture-laden to disperse like the cloud from an air burst, the Baker cloud would drop its fallout as radioactive rain within a few thousand yards of the detonation point. What the experts failed to foresee was how the expelled water would return to the lagoon. Although the exact ratio remained unclear, only about half fell as rain. The base surge accounted for the other half, and for much the larger share of contamination. Neither expected nor understood, the moving blanket of radioactive mist deposited active products on the target ships in amounts far larger than had been foreseen.[63] Obscure though the reasons remained at the time, the consequences stood forth clearly enough. In his final Baker day dispatch to the joint chiefs, Blandy noted what it meant: "Detailed examination of target ships may be delayed several days by radioactivity persisting in water and on board."[64]

From a cautious distance, the target fleet appeared largely intact. The vessel that held the bomb had vanished; a battleship and an oiler sank almost at once; a landing craft capsized and went down several days later. Heavily damaged and far too radioactive for any salvage attempt, the aircraft carrier *Saratoga* flooded and sank that afternoon. Only four other ships showed much sign of trouble as Baker day drew to a close: two battleships, a destroyer, and a transport. The toll mounted, however, as high levels of radioactivity hampered salvage efforts. Progressive flooding sent a second battleship to the bottom on 29 July; a floating drydock followed for the same reason on 6 August. Both destroyer and transport required beaching, as did a damaged submarine. Four other submarines, submerged for the test, went to the bottom but salvage units managed to raise two of them. The final count stood at nine vessels lost and three beached to prevent sinking. Realistically, however, the rest of the target fleet merely stayed afloat; contamination remained so high as to frustrate all efforts to restore most ships to working order.[65] As late as 9 August only ten target vessels had been "declared 'Geiger Sweet' by Initial Boarding Teams (below .1 roentgen per 24 hours)."[66]

That could not be known in Baker's immediate aftermath, but clearly "the danger zone was sharp & high in dosage," Stafford Warren wrote home the day after the test. "I took the press & Adm Blandy & [Gen.] Nichols thru the target area today in a PGM 31 20 ft gunboat. We scared the press & in ½ hr got just a tolerance dose."[67] For three days after the bomb exploded beneath Bikini lagoon, Joint Task Force One could do little but watch and wait. Blandy's daily messages to Washington stressed the same theme again and again. "The target array continues . . . radiologically hot and prevents boarding of ships," he wired on Baker plus one.[68] "Dangerous radioactivity persists throughout entire area of target array" was the next day's report.[69] A day later "intensity of radioactivity near center of array has diminished but still is such as to prevent long continued salvage or inspection activity on any target."[70]

Priority during the week after Baker went to keeping ships afloat and recovering test data, especially the animals aboard for the test, 200 rats and 20 pigs divided among four ships. Five days passed before salvage teams managed to retrieve the last animals. Persistent radiation had been a small problem in the first test, and all animals had been removed within two days.[71] Longer exposure meant higher mortality rates among Baker than among Able animals, but radiation was the killer.

> Although the animals were all situated in interior rooms on vessels located upwind from the Zeropoint, the great majority of them had died by 1 Nov 46. In nearly all cases, cause of death was gamma radiation. Dosages received varied from 310 roentgens (BRACKEN, at 1420 yd) to 2700 roentgens (GASCONADE, at 580 yd).[72]

No pigs, in fact, survived a month, and only a third of the rats.[73]

DECONTAMINATION

Other attempts to reboard target ships at Bikini were limited for days. "The high residual radiation intensities remaining on and in most of the target ships presents an extremely difficult and dangerous problem," warned Stafford Warren.[74] Allowing crews to return seemed at best a distant prospect, but four days after Baker the pace of salvage work began to increase under orders of Director of Ship Material Solberg. Decontamination commanded the strongest effort. Traditional navy techniques—hosing ships with seawater, scraping decks and bulkheads—met only modest success. Rad-safe control aboard the *Haven* observed on 5 August "that the decrease in . . . radiation on ships . . . in the process of decontamination is not on the average significantly greater than on ships not being decontaminated."[75]

Initial washing might reduce a ship's activity by half, but that still left most ships unsafe except for brief visits. Tolerance, in fact, was usually expressed in minutes; monitors sectioned ships by the time required to receive a tolerance dose. An hour or less remained the common limit for a day's work in many areas.[76] Experimental approaches using special chemicals fared no better. Only stripping exposed surfaces by sandblasting or strong acids enjoyed much success, but such techniques were themselves dangerous and ill suited to large-scale use. The navy persisted in its efforts because there seemed no choice. Extrapolating from the readings of 3 August, natural decay would not reach a safe level—the accepted 0.1 roentgen per day standard—on many ships for weeks, on some for months, on the worst until September 1951.[77]

Radiological safety problems extended beyond the target fleet. Nearly a week elapsed before Blandy allowed all ships back in the lagoon. Contamination had by then already forced one change of anchorage. "Radioactivity in lagoon is showing considerable persistence," Blandy informed the joint chiefs on 27 July.[78] "Enlargement of contaminated area to south and east," he reported the next day, "has forced to sea or to less desirable area to westward the ships which had been brought in."[79] The trouble proved to be only temporary: "We are over the hump now," Stafford Warren optimistically wrote on 30 July, "the fleet back in the lagoon again (on the 5th day as predicted) with the radioactivity just below tolerance everywhere except on the target ships."[80]

Radioactivity also began spreading to support ships, but at first it aroused little concern. "Complications involving the operating fleet . . . such as the radioactivity accumulating on the marine growth on the hulls, in evaporators, on the beaches, in the boats, and from being tracked in on the decks of ships themselves, etc., etc.," Warren dismissed as "minor, . . . a nuisance."[81] Contaminated work clothing and gear explained part of the problem. "Personnel decontamination"— procedures workers were to follow, what they should wear, how they should bathe, when and how their clothing should be laundered, when dumped at sea—became the subject of new rad-safe rules.[82] Designating one ship as change station and laundry also helped, but the problem was never fully solved.[83] Other problems proved even more recalcitrant. Ships used large amounts of seawater for routine purposes, and water lines began causing Geiger counters to chatter. Evaporators that distilled seawater for drinking also concentrated fission products and produced even higher readings. Although foreseen and in part prepared for, the reality proved more troublesome than had been expected.[84]

Perhaps the least expected threat came from marine life that showed a talent for concentrating active substances from the water. Algae growing on the outer surfaces of ships' hulls proved all too easy to detect within.[85] Monitors on one destroyer took readings of 0.156 roentgen per day at some outboard bunks on 28 July and 0.204 roentgen per day at the ship's evaporators.[86] Assured "your present situation not dangerous," the ship was ordered to sea for flushing with clean seawater.[87] Generally, flushing water lines and scraping algae from hulls proved helpful in holding down readings. Keeping ships away from troubled waters served best, but that was not always an option. "Essential functions," Solberg reported, unfortunately demanded "a good many exceptions." Salvage vessels, rad-safe patrol boats, and scientific survey craft all needed access to contaminated waters: "The only immediate solution on these vessels lay in constant monitoring, restriction of hazardous areas, and scheduling of operations wherever possible in a manner designed to avoid further increase in radioactivity in portions of the ship already at dangerous levels."[88]

The rad-safe section formed a group to survey task force ships. "Contamination was found on all of the ships visited," it reported, but "except for a few isolated cases, no physical hazard could be expected." Permanent monitors were assigned to a few ships, bad practices were corrected, and "a number of sleeping quarters were evacuated (at our request) because they were slightly above tolerance (> 0.1r/day). This condition was due in every case to the proximity of salt water pipes or to the sides of the underwater portion of the ship."[89] Blandy shifted anchorage again early in August, this time to the lagoon's entrance. "The job here is very strenuous & we are pushed for time," Stafford Warren wrote his wife. "We moved the fleet today to cleaner water. . . . The radioactivity is a more serious thing than they tho[ugh]t."[90]

Accelerated salvage work from 29 July onward increased the number of high film-badge readings, sometimes pushing the film's upper limit. Strenuous efforts nonetheless kept most readings below tolerance. From Baker day until it departed with the rest of the rad-safe section in mid-August, the photometry group distributed 6,000 film badges; as with Able, only half went to personnel. The group recorded overexposures from less than 10 percent of the total, most under 0.2 roentgen.[91] Unfortunately, this record told only part of the story. "Increasing evidence of exposure considerably over the tolerance limit" as early as 3 August concerned Stafford Warren:

> Certain key personnel have already accidently received as much as 2 R. in two days' time. Several other key personnel as well as enlisted men have received as much as 0.3–1.0 R. on separate days 2 or 3 times. While this probably has not seriously harmed them it is an indication of what will continue to happen in spite of all our precautions and the care of these men themselves to avoid over exposure. The increased number of personnel going aboard or alongside radioactive target ships, the wide variation of radiation intensity at different locations on the targets ships, and the impossibility of continually

keeping track of all men on board are probably the main causes of this. It is anticipated that these problems will increase rather than decrease as time goes on.[92]

To make matters worse, Warren added, "There is very little leeway in this since approximately 10% of the daily permitted dose is already taken up by the exposure to which all the Task Force are subjected every day while living in the lagoon."[93]

Film badges were of little use for beta radiation, which proved a larger problem than expected. The rad-safe medico-legal board concluded that "the permissible dose of beta irradiation to the entire body will not be exceeded if the gamma irradiation is held below 0.1 r per 24 hrs. (This is on the basis of limit of 0.5 r of beta per 24 hrs.)"[94] Beta-gamma ratios, however, varied widely, and whole-body exposure was not the chief danger. "Contamination of hands and faces with beta emitters greater than tolerance," Warren observed,

> is exceedingly common. It is not infrequent to find personnel with amounts on the bare hands bordering on erythema dose levels (if not removed within 24 hours). It is almost impossible to enforce the wearing of gloves continuously on badly contaminated ships, during the clean up stages under present circumstances where large numbers of men are involved. Nor is it feasible to expect them to take the proper care of their contaminated clothes.[95]

Enforcing rules depended on good monitoring, itself becoming a problem. The rad-safe section was losing its trained personnel. "Monitoring demands have been increasing steadily while our numbers are being depleted," Warren observed.[96] "The Radiological Safety Section . . . will lose about 350 expert monitors and instrument men who will return by air or ship by August 15th." With no one to replace them promptly, "this leaves a small handful of approximately 24 Navy and Army men, and a few civilians recently trained as monitors to protect the large body of men working on the 70 odd target vessels."[97] Warren's forecast proved pessimistic—the section retained seventy-two officers and sixteen enlisted men when the rest departed in mid-August[98]—but the higher number, too, fell far short of need. Instruments presented yet another problem area. Monitors complained of erratic readings, of devices that too often failed to work at all. Replacements arrived slowly, if at all, and spare parts often were lacking even when repairs might otherwise have been feasible.[99]

His review of the safety situation on 3 August convinced Warren that too much contamination and too few monitors posed real danger to the task force. He urged a quick end to Operation Crossroads: "The remaining time between now and 14 August should be spent working on such vessels as have relatively little radioactive contamination or where the usefulness to the Task Force is great and the effort and rush is worthwhile." All other target ships "should be declared hopelessly contaminated and be towed to shallow water and beached and time allowed for radioactive decay to take place."[100] Warren's medico-legal board found itself "in complete accord" with these recommendations.[101] Solberg and his salvage crews viewed matters differently. Although needing "the most elaborate precautions . . . [and] constant vigilance to insure that all personnel were safeguarded adequately," insisted the director of ship material, "the work proceeded very satisfactorily."[102] Joint Task Force One's high command reacted cautiously. Decontamination efforts would be restricted, they decided in meetings on 5 and 6 August, but not as severely as Warren proposed.[103]

Warren, however, would not let the matter rest, and the next day he renewed his case with further arguments. Declining gamma readings brought "more and more pressure . . . for putting full crews aboard to work and live and eat aboard."[104] Privately, Warren also revealed that he had begun "to get proof that the boarding & 'cleanup' parties were getting over the line & avoiding monitors etc. Since they couldn't taste, feel, see etc. anything, the officers & men began to take advantage of their numbers & my green men."[105] Lyon later decried the officers who adopted a "blind, 'hairy-chested' approach to the matter with a disdain for the unseen hazard."[106] Scientists eager to obtain data might also neglect safeguards. Reporting to Los Alamos, one observed "it would perhaps be judicious if Col. Warren were not informed of the origin of this sample; boarding the Nevada was verboten at the time the samples were taken, and our men must have taken [approximately] 1 R in getting it." Decontamination, he also commented, seemed a problem only if "it is insisted that no man should take more than 0.1 R per 24 hours while working on these ships, . . . a preposterous limitation."[107]

Monitors might well believe they detected "an attitude of indifference on the part of the ship's officers . . . to the safety standard set by RadSafe." The two assigned to the *Prinz Eugen* thought so: "There is reason to believe that men are being kept aboard for longer periods of time than they should be and also that the ship's officers believe that the standard of .1 R has such a large safety factor that it can be ignored."[108] Criticism, however, might cut both ways. "There are as many different readings as there are monitors involved in their taking, and this factor must be reduced to uniformity in order to ascertain accurately progress being made with methods used," complained the *Prinz Eugen*'s commanding officer. Instruments failed to meet navy standards, and he thought "all hands engaged in decontamination work should be provided with gages of some type to be examined on their persons and to serve as a warning as to the exposure they have received, this for morale as well as for safety."[109]

As gamma readings declined, crews worked longer shifts or moved aboard their ships, raising for the first time the hazard of alpha contamination. Unfissioned plutonium, like fission products, had blanketed the target fleet, or so at least the radiological safety section believed.

> The unreacted components of the bomb were deposited in the same way as the fission products so that wherever fission products were detected the former may be assumed to have been present also. However, experiments on some ships showed that washing procedures might remove fission products without removing the unreacted components. The presence of these components was determined analytically in above tolerance concentrations in samples collected from a number of target ships.[110]

Scientists from Los Alamos working on Kwajalein analyzed samples collected from the target vessels. Their findings seemed to confirm not only outer surfaces but "inner compartments of all target ships highly contaminated by alphas."[111]

Plutonium, if actually present, would pose serious rad-safe problems. Unreliable techniques for monitoring alpha contamination depended too heavily on guesswork and supposition.

> Decontamination requires meticulous care and an elaborate set up of equipment and trained men, none of which are available within the resources of the JTF-1 and the Manhattan District. Trained personnel and equipment will have to be developed over a matter of four or more months of intensive work and instrument building.[112]

Later studies would show that the findings of early August 1946 were misleading: inadequate instruments and misguided judgment painted a danger far greater than actually existed. Warren, of course, knew nothing of future findings. On 7 August he again urged a halt to work even more strongly than he had four days earlier. Target vessels, he observed, are

> extensively contaminated with dangerous amounts of radioactivity. Quick decontamination without exposing personnel seriously to radiation is not possible under the present circumstances and with present knowledge. Control of the safety of target ship's crews is rapidly getting out of hand. Adequate monitoring personnel and instruments are no longer available.

Joint Task Force One, Warren concluded, faced "great risks of harm to personnel engaged in decontamination and survey work unless such work ceases within the very near future."[113]

Matters came to a head on 10 August, as Blandy and Parsons prepared to sail for Hawaii. In a letter to his wife the next day, an elated Warren described "the battle won in all details—namely to stop the program as it is now." He went to the farewell luncheon with

> analyses just in by dispatch from Los Alamos in the nick of time to nail it down & when Parsons saw it, he said, well this stops us cold alright & supported me fully. The table was cleared at 1:00 pm & we got down to cases. They were to sail at 6 pm. Adm. Solberg said he was convinced, & it was his group of ab[ou]t 2000 men whom I was fighting—he was swell but hard to convince because he felt he could clean up these ships & I'd been saying he was butting his head against a stonewall & was only fooling himself & risking a lot of men. A self x ray of a fish & 3 tables & a curve did the trick. Blandy said, "If that is it, then we call it all to a halt." He called in the staff & worked like mad. Dispatches by the score, reports changed, etc. . . . Solberg waited as I finished up the final details & we both stept off the gangplank saluted & they pulled the gangway up & sailed—as close as that!![114]

Rear Admiral F. G. Fahrion, commander of the target vessel group, took charge of the so-called stabilization period at Bikini when Blandy left. He fully supported Warren's repeated demand that "no further work must be permitted in the contaminated target ships."[115]

THE END OF CROSSROADS

Remaining units of Joint Task Force One abandoned Bikini in mid-August, shifting base to Kwajalein. As Blandy explained,

> Although all waters and land areas of Bikini atoll are well within radiological tolerance, the tendency of radioactivity to concentrate and

accumulate in ships, especially in evaporators and in marine growth on the hulls, makes it mandatory to remove the ships of the task force from this atoll with its small and decreasing but nevertheless cumulative hazard.[116]

It was none too soon for Warren. "I'm getting the jitters," he wrote his wife just before himself leaving Bikini, "& so are the rest because there are so many men here altho they are leaving in large no's daily."[117]

Construction for Test Charlie continued at the west end of Bikini lagoon. That deepwater test, however, had never been firmly scheduled. Unexpected problems in the aftermath of Baker may have exacerbated other doubts. Cancellation, in any event, was widely rumored even before Blandy sailed for Pearl Harbor on 10 August. After a week of talks at the navy's Pacific headquarters, he hauled down his flag from the *Mt. McKinley* and flew to Washington.[118] On 7 September President Truman advised Secretary of the Navy Forrestal that he had "decided to indefinitely postpone this project."[119] A public statement was issued the next day. Project Crossroads had ended, but Joint Task Force One still had work to do.

Many ships departed Bikini before the full extent of the problems became clear. Each left with radiological clearance and a tailored list of safety rules, mainly limits on the time crew members might spend in certain parts of the ship. That no longer seemed enough by mid-August. Every ship at Bikini during or after Baker day received new orders: monitoring must precede dry-docking or repairs.[120] These measures, too, fell short of needs. Warren and Lyon made further suggestions, and Captain W. E. Walsh as newly appointed force medical officer established a ship-clearance organization in San Francisco. Monitors were scarce, however, proper techniques still posed questions, and the size of the problem remained unclear although it seemed to grow ever larger.[121] Recommended measures "were, at best, partially effective and grossly inadequate to meet the problem," Solberg reported.[122]

The navy meanwhile outlined a service-wide radiological safety program late in August. In simplest terms, the Bureau of Medicine and Surgery would set standards, the Bureau of Ships would apply them, and the Bureau of Personnel would provide training.[123] Work was soon under way. Decontamination experiments at San Francisco addressed one problem. Before year's end this program took institutional form on the way to becoming the Naval Radiological Defense Laboratory.[124]

Experimental results came quickly and, combined with experience from Bikini, furnished Solberg the basis for a decontamination plan. It was issued later that month under the joint auspices of the Bureau of Ships and the Bureau of Medicine and Surgery. Authorizing work to begin and telling how to do it, the new directive opened the navy's concerted effort at coming to terms with the aftermath of Crossroads.[125]

Temporary assignment boosted the number of monitors to cope with the other major problem. Solberg's plan also made monitors

> desirable but not repeat not essential. . . . Tests and development work have demonstrated that all of the above specified procedures can be carried out safely and without exposing personnel to any hazards. However in order to insure absolute safety personnel should be supervised and precautions . . . observed.[126]

To meet longer-term needs, the navy founded a radiological safety school. Early in August Blandy had asked Lyon to begin an emergency training program at Bikini. The Navy Department called Lyon and his staff to Washington. Instruction actually began for selected members of the task force en route from Bikini.[127] Formal schooling—four weeks of classroom work followed by three months of field training at San Francisco and Kwajalein—commenced in Washington on 9 September. A second course followed in November.[128]

Questions about standards for clearing ships took longest to resolve. Experimental findings, panels of experts, issues of publicity and security, even some guesswork, affected the final decisions.[129] As the Manhattan Project's expert on internal emitters, Berkeley physiologist Joseph Hamilton, commented in one meeting, "At present the establishment of standards is mainly guess work and calculated risk . . . , clearance is mainly a case of calculated risk and good judgment."[130] Eventually the navy defined two clearance levels: operational and final. Operational clearance meant crews could work hazard-free using standard radiological safety precautions. It required readings no higher than 0.5 roentgen per day, gamma and beta combined, anywhere on the ship. Readings from metal-enclosed saltwater systems could not exceed 0.1 roentgen per day gamma. Anything higher demanded prompt decontamination or disposal. Corresponding figures for final clearance were a thousand times lower: unshielded 0.005 roentgen per day gamma and beta combined, shielded 0.001 roentgen per day gamma. Higher than natural background radiation, such levels none-theless were deemed to pose no radiological hazard of any kind.[131]

Although expressed as gamma or combined gamma-beta readings, clearance standards really concerned alpha contamination. "All of the ships involved," the navy noted, "have low radiation intensities and small amounts of contaminating materials. They present no danger from external radiation. Any danger to personnel which may exist in-volves the introduction of contaminating toxic materials into the body."[132] Expedience dictated the seeming anomaly: instruments for reading gamma and beta in the field existed, whereas alpha detection and measurement remained largely a laboratory project; empirically de-rived ratios, however, allowed more easily detected beta-gamma to serve as an index for hard-to-find alpha.[133] "Monitoring is just a rough indication of what we are really after—the alpha rays."[134] To direct the clearance effort, the navy formed a radiological monitoring organiza-tion at San Francisco. The Bureau of Ships and the Bureau of Medicine and Surgery issued a joint directive on 22 November 1946, listing gen-eral safety precautions and detailed decontamination procedures. De-contamination efforts now extended to every task force ship that might have been exposed at Bikini, whether destined for the active fleet, mothballs, or scrap.[135]

Much work had already been done. At a late-November meeting in Washington, Solberg announced "that 90% of the vessels are expected to have final radiological clearance by 20 December."[136] That same meet-ing discussed and approved new, less stringent rules for sand and acid, the wastes of decontamination. Sand used in sandblasting no longer required special handling and disposal, and diluted acid used in clean-ing saltwater systems might now routinely be discharged into har-bors.[137] Hamilton had no doubt "that the quantities involved entail ab-solutely no health or security hazard." Nevertheless, fearing the public's "likely failure to understand that no hazard from radioactive materials existed," he urged "consideration . . . to the public relations angle in not permitting the information to leak out regarding the local disposal of acid and sand containing some fission products."[138] Ham-ilton likewise affirmed that scrap metal from Crossroads ships posed no danger. For that reason he dismissed any possibility that someone who received such scrap might win a suit against the government. To General Groves's fear "of claims being instituted by men who participated in the Bikini tests," Hamilton replied in the same vein: "There is absolutely no possibility of physical injury from radioactive materials in the amounts which are being worked with on the non-targets under present conditions."[139]

By late October no ships remained under the operational control of Joint Task Force One. On 1 November 1946 the force was formally dissolved, although a joint crossroads committee under Parsons survived to collate data and complete reports. The best efforts of the salvage teams in the end had allowed only eleven target ships to sail from Bikini under their own power. Decontamination accounted for but six of that number, five submarines and a destroyer. The other five—transports from the fringe of the target array—were neither damaged nor contaminated on Baker day. The rest of the survivors, still too active to permit crews aboard, required towing. At the urging of Warren and Lyon, the navy postponed plans to return target vessels to the United States, towing them first only as far as Kwajalein.[140] "Radioactive contamination . . . even more severe than anticipated" had convinced Blandy "that it would not be practicable to decontaminate the bulk of the surviving targets sufficiently to permit their early return to the U.S."[141] Uncertainty about the fate of these ships in fact persisted for weeks, and most of them never returned home. Meanwhile they remained at Kwajalein, off limits except by direct order.[142]

Unloading munitions was the major task at Kwajalein. "All safety precautions, including radiological, are being scrupulously observed," Admiral Fahrion reported, but unstable munitions in contaminated ships made for "a slow job."

> The physical discomfort incident to wearing full clothing and rescue breathing apparatus is aggravated by prevailing tropical heat and humidity. In addition, men wearing rescue breathing apparatus find it difficult to see, due to restricted angle of vision and fogging up of face pieces by moisture and perspiration.[143]

Such "extreme precautions," as they seemed to the officer in charge of the ammunition disposal unit, "against possible, but actually insignificant, concentrations of radioactive material in the air led to the amplification of explosive and other types of physical hazards."[144] Despite stringent measures to preclude overexposure to external radiation, work crews on the two-month job often recorded film-badge readings slightly above the allowed daily limit of 0.1 roentgen.[145]

In the face of persistent radioactivity, decontamination of former target vessels had long since faded as a goal. The joint crossroads committee had reviewed the readings taken during the two months after Baker and concluded that

> all of these ships are unsafe for unrestricted clearance. Since some of the active material is decaying at a very slow rate, since these ships have already been extensively decontaminated, and since the contamination is by no means localized and probably covers at least the entire topside surface, it seems improbable that they could ever be decontaminated to the point where they would be safe for use, even for scrap purposes.[146]

These data as well as later readings convinced an unduly pessimistic Stafford Warren that "it would appear hopeless to attempt any further large scale decontamination on any but an experimental basis." Like the joint crossroads committee, he recommended sinking all the target ships at sea, except for a few "of definite experimental or training value to the Navy."[147] The navy decided to tow twelve former target vessels to the mainland for more detailed study. Six were destined for the naval base at Bremerton in Puget Sound, six for Hunter's Point in San Francisco Bay, where they also became training grounds for future monitors. The ships received a careful examination when they arrived, and work crews were instructed to behave with the utmost caution. The navy insisted on the full range of established safety measures, from frequent medical examinations to the regular use of special clothing, gas masks, and film badges.[148]

The radiological aftermath of Crossroads deeply impressed thoughtful persons. It underscored, Parsons observed, "the deadliness and awesomeness of the Atomic Bomb."[149] Whether exploded in the air or underwater, the bomb could destroy a fleet. Most ships might survive, but exposed crew members would sicken and die. Combined data from the two Crossroads tests painted a clear picture. Radiation damaged animals much as it damaged people, although species varied somewhat in susceptibility. Deposited fission products from the underwater bomb created a greater hazard than the burst of radiation from a bomb exploded in air, and even heavy armor offered no certain safety. Estimated lethal range of the Able airburst was less than 1,300 yards in the open, half that for crewmen behind armor. Instantaneous exposure from the burst itself would have caused all the damage.[150]

Had crews "been aboard Test 'A' target vessels" judged the joint chiefs evaluation board, "measurements of radiation intensity and a study of the animals exposed in ships show that the initial flash of . . . gamma-rays and neutrons would have killed almost all personnel . . . stationed aboard the ships centered around the air burst and many others at greater distances."[151] Baker produced a different pattern. Contamination extended more than a mile upwind, well above twice that downwind, reported the task force historian. Cumulative rather than instant exposure became the chief threat. Crews would have fared no better in Baker than in Able, although not for the same reasons.

> Topside personnel within 700 yd would have received lethal dosages (400 roentgens) within 30 sec to 1 min and would have received roughly 20 times the lethal dosage (8000 roentgens) within the first hour; personnel within 1700 yd would have received lethal doses within 7 min, and those within 2500 yd (crosswind or downwind) would have received lethal doses within 3 hrs.[152]

In the more vivid terms of the joint chiefs evaluation board, "the second bomb threw large masses of highly radioactive water onto the decks and into the hulls of vessels. These contaminated ships became radioactive stoves, and would have burned all living things aboard them with invisible and painless but deadly radiation."[153] Whether exploded above or below the water's surface, a bomb would leave few crew members within a mile unharmed. More distant crews would suffer less, but the death toll near and far would mount as days and weeks passed.[154] When asked at a press conference what safety measures might cope with such prospects, Blandy's answer was brief: "Principally, removal of the patient before exposure. The best protection is not to be there when the bomb goes off."[155]

In this way, in effect, Joint Task Force One protected itself. Radiation killed pigs, goats, and rodents, but no people. Fission products from Baker remained trapped in Bikini lagoon, spreading through every form of life that ecosystem supported, but again not to people. Contaminated ships baffled the best salvage efforts of the task force, but careful work and strictly enforced rules largely kept exposures to workers and crews within safe limits.[156] Later findings would alter the picture, but at the time it looked like a close call. Stafford Warren suggested as much in a letter to his wife just before he left Bikini. "We have pulled out of here the majority of the force . . . just in time."[157] Lyon agreed, as he reported to the surgeon general of the navy:

> We felt it was definitely unwise to continue on the large scale operation we were conducting, in view of the relatively few radiological

safety monitors and instrument experts we would have available after the 14th of August when the bulk of the Radiological Safety Section had to return to the States. . . . However, I am certain that we ceased major operations during the period of real safety, and that we did not extend over into the period of serious hazard. . . . [B]oth Colonel Warren and I feel we have stopped the large scale operation short of the danger point.[158]

One monitor thought so, too—"probably no permanent radiation injury was sustained by any of the participants"—but he also expressed some doubts to Warren: "I do believe . . . many of us probably received much more penetrating ionizing radiation than the instruments . . . were able to record."[159] Warren demurred only in part, confessing, "I never want to go through the experience of the last three weeks of August again. The air inhalation possibilities and all of the rest indicated conclusively that, just upon the basis of statistics alone, we were certain to get into trouble if we did not close the operations shortly."[160]

Such views, valid or not, remained unknown to the public. The official version more closely echoed the rad-safe medico-legal board: "Tests Able and Baker were carried through without irradiation injury to any persons."[161] Despite problems far more troublesome than expected, "there have been no casualties from overexposure to radiation," Blandy concluded. "No man was so exposed as to give rise to apprehension that he might become a casualty at a later date. In fact, there is no evidence that any person suffered any ill effects whatever from either of the two atomic bomb explosions at Bikini."[162] Stronger doubts have since emerged. Officials now stress intent and context rather than accomplishment. "All CROSSROADS operations were undertaken under radiological supervision intended to keep personnel from being exposed to more than 0.1 roentgen (R) per day," comments the Nuclear Test Personnel Review. "At the time, this was considered to be an amount of radiation that could be tolerated for long periods without any harmful effects on health."[163] Others have questioned even so cautious a judgment.[164] All such views are long after the event. Apprehension existed in 1946, but its focus was future war, not the health effects of tests at Bikini.

Epilogue

CONTINUITY AND CHANGE IN RADIATION SAFETY

The dying echoes of Crossroads Baker marked more than the end of a test. Operation Crossroads looked very much like World War II's final campaign, though mounted almost a year late. A joint task force modeled on those that conquered Japan's island strongholds conducted the tests at Bikini. The same army-directed team that developed atomic bombs during the war furnished the bombs and experts for the postwar tests. Yet neither Joint Task Force One nor Manhattan Engineer District outlasted 1946. With its mission completed the task force dissolved, and the army project gave way to the civilian Atomic Energy Commission. Testing of nuclear weapons assumed quite another guise under its new sponsor, but radiation safety remained as serious a matter of concern.

Ionizing radiation from natural sources has always formed part of the human environment, sometimes with harmful effect. Mysteriously high cancer rates long known to afflict certain miners, for instance, proved the result of breathing radioactive gases underground. Tracing that cause, however, became possible only when science revealed X rays and radioactivity in the nineteenth century's closing years. From these discoveries also dated ionizing radiation as a man-made hazard. Medical and commercial uses of X-ray machines and radium harmed as well as helped. Multiplying injuries, in turn, spurred efforts to devise safeguards against and impose limits on radiation exposure.

For a long time such limits remained chiefly self-imposed; personal judgment and action defined the only controls that existed until the 1920s. Then widely reported injuries and deaths among those working professionally with X rays and radium aroused public concern. So, too, did the unhappy fate of unwitting victims of radium poisoning, like the young women who painted luminous figures on watch dials. Professional users acting in concert through such groups as the American Advisory Committee on X-Ray and Radium Protection and the International X-Ray and Radium Protection Committee responded to the public outcry and to their own concerns. The self-appointed wardens of radiation safety had by 1941 published standards for exposure to X rays and gamma rays, for body burdens of radium, and for air concentrations of radon. Rarely bearing the force of law, such standards remained more guideline than rule. They nonetheless won broad support and helped reduce the toll of injuries. They also came just in time to meet a vastly larger need.

The advent of controlled nuclear fission and then atomic bombs during World War II transformed the nature and scope of radiation hazards. Nuclear energy exploited for war required huge new plants of novel design and unprecedented kinds of field-testing. These activities not only put more workers at risk than any prewar program but might also threaten large numbers of people merely living nearby. Prewar safety standards and practices nonetheless met most wartime needs. Radiation safety in the Manhattan Project remained firmly based on the methods proved through nearly five decades of trial and error. Developing nuclear reactors and bombs might mandate new safety measures ranging from routine use of film badges to founding health physics, but such changes, for the most part, simply augmented well-tried techniques. More extensive changes marked the postwar world. Radiation safety faced new demands (as we shall see in a later volume), political as well as technical. In 1945 and 1946 that all remained largely in the future.

Yet many questions raised before and during World War II presaged the concerns of the 1950s and 1980s. Commonly cast as questions for science, the real debate more often involved philosophy and public policy. Social concerns in the widest sense always molded safety standards, science at best setting guidelines for decision makers. The key issue was, and remains: How much radiation exposure should a worker or a member of the public be allowed? Controversial from the outset, this question nonetheless had, in one sense, an easy answer: not enough to cause harm. Unfortunately, the easy answer simply translated the first question into a second: How much is harmful? And that question, after almost a century of research, has yet to receive a final answer. Radiation safety in laboratory and plant, however, could proceed with less. Uncertainty never precluded standards being proposed and set, rules announced and observed, people normally protected against the worst hazards of overexposure.

Ionizing radiation can damage biological systems. About that no doubt has existed from the very early days of X-ray use. Practical experience and laboratory findings alike confirmed that exposure to radiation might endanger health and life. Radiation, however, also conferred benefits. Medical practice and research became only the first of many areas transformed by the new tools available after 1895. Thus the heart of the problem was, then as now, the hardest question: When might benefits outweigh risks? But phrasing so simple masked deeper ambiguities. In the years before World War II, as since, both "risk" and "benefit" proved words of protean meaning.

Ambiguity began with the study of radiation-caused harm to living things. Radiation damage depended both on the amount absorbed (dose) and on the rate of absorption (dose rate). A dose that was lethal if received in a day might well be survived if spread over a month and prove all but harmless if stretched through years. Precisely what physical processes converted absorbed radiation into specific kinds of biological damage were then, and still are, unclear. Inevitable death followed only very high acute doses. Exposure at lower levels and rates produced much less clear-cut results. No one could say a certain dose would cause a certain injury to a certain living thing. Prognosis became an exercise in probability, as witness the so-called median lethal dose (LD_{50}): the dose sure to kill half a large number of exposed subjects within some specified time. Statistically, one might know that half of those exposed in twenty-four hours to 400 roentgens would die within thirty days, yet remain helpless to predict which persons that half would comprise. Similarly, knowing what ratio of those who inhaled radon at a defined level would surely develop lung cancer allowed no one to foretell specifically who among them would succumb. The tension between statistical group effects and individual harm was, however, only one factor in a complex problem.

Acute exposure produced effects clear enough for prompt action. Straightforward measures like shielding sufficed to safeguard the few workers at risk. Exposure at lower levels, however, posed harder questions: larger numbers of people threatened might offset lesser risks to any single person. Controversy centered on just this issue: the effects of exposure to low-level radiation. But what to call "low" changed over the years. No new danger needed proving to invest exposures regarded calmly in one decade with deep concern in another. Technology was a major reason. Improving instruments detected ever lower levels of radiation. As for many other toxic hazards, detection in practice if not in theory implied danger. Technical prowess, in other words, rather than assured hazard tended to define safe limits. This tendency itself became a problem during the war as work moved out of the laboratory. Instruments suited to expert hands at the bench proved too often unreliable, tricky to use, and hard to interpret in the field. That demand normally exceeded supply merely compounded the problem.

During the war the key questions still centered on immediate or acute or short-term effects versus delayed or chronic or long-term effects. Evidence of damage appeared more slowly and numbers injured declined after dose and dose rate began falling. Someone exposed to a massive burst of gamma rays quickly showed the effects, leaving no doubt about the cause. Exposure to some lesser dose might induce leukemia or other disease years later, when the cause became far less clear. At still lower levels and still longer times between exposure and seeming injury, causal links grew ever fainter. Medical and legal questions alike multiplied as ties between cause and effect stretched. Just how very low dose triggered biological response remained obscure. Natural background radiation also varied enough, for such reasons as geography or altitude, to mask small effects. Scientific opinion thus divided about the shape of the dose-response curve at very low levels where cause and effect grew most difficult to measure or even to detect.

Scientists taking one approach in effect graphed the curve as a straight line from known through unknown dose-response values to zero dose. Any exposure thus implied some chance of harm, even if damage could not always be detected. Interpolating linearly, in other words, meant that only zero dose caused zero damage. Another approach framed the problem in more familiar terms. Biologically active agents, such as drugs or poisons, normally must exceed some threshold before working. Biological systems exposed at levels below threshold could restore themselves and so suffer no lasting effect. Radiation at low enough levels, should this analogy prove valid, caused no cumulative damage. The well-known fact that doses that killed when acute only injured when chronic seemed to support the threshold theory. Interpolation suggested that long-term effects from low-level exposure likewise tended to vanish. The dose-response graph, in short, might turn sharply downward toward zero damage at some dose higher than zero. This viewpoint largely prevailed during the first half century of radiation safety.

Experiment could not easily resolve the issue. Meaningful data on the rare and often minor damage inflicted by very low levels of radiation could come only from huge numbers of animals studied over long periods of time. Possible in theory, such studies simply exceeded the limits of any realistic research program, especially because animal findings might leave unanswered too many questions about human effects. Practically, that presented no insoluble problem. Radiation safety never relied on final answers. Pragmatic safeguards countered the everyday hazards long before science could explain either hazard or safeguard. Threshold thinking shaped early safety codes. "Tolerance" expressed the basic idea: living things could survive without patent ill effect some defined level of radiation for an indefinitely long time. Inhabitants of Denver, after all, seemed as healthy as New Yorkers, although Denver's altitude meant they received in added background radiation half again or more than dwellers at sea level.

Dominating safety thought through the 1940s and beyond, the threshold approach underlay all Manhattan Project safety programs, though not without challenge. "Permissible exposure" emerged as an alternative concept in the mid-1930s. The newer term added social-political views about what might be allowed to medical-biological judgments about what might be harmless. Radiation safety, in effect, shifted from seeking biological thresholds to weighing risks and benefits. Although the weight of evidence in fact argued threshold, guideline writers assumed the philosophical stance that any exposure was risky. Whatever they believed, experts came to prefer erring on the side of caution, acting as if any exposure could be harmful.

That approach to radiation safety fully triumphed only years later, despite growing support in the Chicago Health Division during World War II. Elsewhere in the Manhattan Project, especially at Los Alamos, safety conceived in terms of tolerance better suited local values. Testing nuclear weapons meant taking risks. Obviously the safest course required no field testing at all, but that option made no sense. Concern instead centered on how to reduce or forestall risks while still getting needed data. Naturally, that goal sometimes eluded testers. Discrepancies between normal standards and actual practice complicated matters. These might be inadvertent: accidents happened, instruments failed, unexpected sources of danger arose. Inexperienced or poorly trained monitors might overlook some danger; uninformed workers or laymen might make mistakes. Familiarity also could breed contempt for any danger, radiological included. In these ways as others, Trinity and Crossroads foreshadowed later problems.

Policy, too, affected safety. Officials set special operational limits for test participants. They also exempted those assigned to certain crucial data-recovery missions from even these higher limits. Special standards, in other words, applied for testing. Deliberation was not always one-sided. Scientists eager to retrieve data willingly exposed themselves to radiation. Chicago's laboratories and workshops might adopt a cautious and conservative approach to radiation safety, but the drama of testing evoked a more flamboyant style at Los Alamos. In some, the setting inspired a kind of bravado that might transcend good sense. Exposure guides subverted unknowingly, in short, posed less a problem than more purposeful behavior by officials convinced that testing nuclear weapons justified modest risks and by individuals willing to risk themselves for reasons of their own. Trinity and Crossroads here, too, set the pattern for later tests.

Tolerance thinking governed the programs intended to safeguard workers and the public during test-firings of the first fission bombs, in New Mexico during 1945, at Bikini a year later. Radiation safety for bomb tests built on well-tried foundations. However much greater the danger from bombs, it differed less in kind than degree from more controlled releases of nuclear energy. Safety planners knew what to expect and how to cope with the dangers. Decisions about testing, however—when, where, under what conditions—took many factors into account. Radiation safety was only one, and not the most important; it might well suffer in conflict with other test goals. Trinity and Crossroads proved such concerns well founded.

Aware that their advice might go unheeded, safety experts ignored no practical safeguard within their powers. They largely succeeded in their purpose. This is not to deny shortcomings: some Trinity workers exceeded exposure guidelines, some bystanders received much higher exposures than anyone would have preferred, some animals suffered beta burns. Yet in a wartime context few doubted the wisdom of the Trinity test. Few, indeed, concerned themselves overmuch with workers and members of the public exposed to what were then understood to be harmless (if perhaps also uncomfortably high) radiation levels.

Overexposures a year later in Crossroads were far lower and fewer. They likewise seemed to matter little, for much the same reasons: no one died, no one suffered any apparent injury.

But times changed, and the past refused to stay buried. We have learned something since the Manhattan Project ended. Applying later standards to the events of 1945 and 1946, some critics have questioned Trinity and Crossroads safety programs, but later standards, as I have suggested, may owe more to technical skill or political climate than to new data or deeper understanding. The weight of evidence, though less overwhelming than it once seemed, still supports judgments at the time that neither Trinity nor Crossroads directly harmed anyone. In World War II America, that was enough. Not now. Professional and public attitudes changed markedly in the next four decades, making long-forgotten details of the 1945–1946 tests matters of renewed concern to Congress and the courts. This new apprehension reflected as much as anything else the safety questions raised by later nuclear weapons testing.

Trinity had proved that implosion bombs worked; Crossroads tested their effects. Proving new weapons and studying their effects became the main goals of later programs. Radiation safety organizations and methods confronted far heavier demands as weapons grew more powerful and testing more frequent. Testing also proceeded under new auspices. With the Atomic Energy Act of 1946, Congress created the Atomic Energy Commission. The new civilian agency formally replaced the Manhattan Engineer District on 1 January 1947. Responsible for peaceful uses of atomic energy, the AEC also controlled bomb design, development, and production through its Division of Military Application. Another division, Biology and Medicine, directed safety studies and programs. The effects of weapons presented distinct, though clearly related, questions outside AEC purview. Responsibility for these problems fell to the newly created Armed Forces Special Weapons Project, working mainly through AEC test programs.

Scientific teams surveyed Bikini in 1947 and periodically thereafter. Radiation remained too high for the Bikinians to come home. Testing resumed in 1948, when a new and much smaller task force returned to the Marshall Islands for Operation Sandstone. This time the site was Eniwetok rather than Bikini, sending another group of natives into exile. The purpose was developmental, the events highly secret. Another long pause ensued. Only in 1951 did the United States embark on the large-scale and frequent testing, now in Nevada as well as the Pacific, which became a public issue. Circumstances changed but programs intended to ensure safety in nuclear weapons testing still confronted much the same problems. In some ways, these problems were present from the beginning.

The central question remained what limits to impose on human exposure to ionizing radiation. It persisted because answers depended as much on social values and philosophy as on research. Research, in any event, failed to resolve all problems. How grave are the long-term somatic and genetic effects of exposure to low levels of radiation, in particular, has resisted any final answer. But not all questions were old ones. Secrecy shrouded the Manhattan Project during the war, and all but one of the bombs exploded a long way from home. Officials had little need to worry about what to tell the public. "Nothing" was the easy answer. That changed sharply after 1951 when the AEC began testing within the continental United States. Educating the public seemed a fine idea, but the line between education and indoctrination proved hard to walk. Public relations became an increasingly central concern in test planning, and public safety, a heated issue. The years after World War II saw new questions raised as well as old questions answered. How the imperatives of testing nuclear weapons collided with growing public concerns about the long-term effects of low-level radiation exposure will be the central theme of another volume.

Appendix

CHRONOLOGICAL INDEX OF RADIATION EXPOSURE STANDARDS

(NOTE: Citations are to chapter and page of the text, where full discussion may be found.)

1914–1924: Introduction and informal acceptance of X-ray protection standards based on the erythema or unit skin dose 1:0–0

1925: Concept of tolerance dose introduced 1:0–00

1925–1928: The roentgen adopted as the international unit of X-ray exposure 1:00

1928: First internationally accepted X-ray protection standard, 1/100 of erythema dose per month 1:00–00

1928–1931: Protection standard redefined in roentgens 1:00–00

1931–1934: U.S. X-ray protection standard recommended, 0.1 roentgen per day 1:00

1934–1937: International recommendation of 0.2 roentgen per day for X-ray exposure limit 1:00

1941: Recommended tolerance dose for ingested radium, 0.1 microcurie residual 1:00–00

1941: Recommended standard for inhaled radon, 10 picocuries per liter of air 1:00–00

1941: Proposed (but not implemented) lowering of U.S. external exposure standard to 0.02 roentgen per day, to counter potential genetic effects 1:00–00

1942: Metallurgical Laboratory, University of Chicago, adopts protection standard of 0.1 roentgen per day for external exposure, which becomes standard for entire Manhattan Project 2:00

1942:   Concept of maximum permissible exposure   2:00–00

1942:   Instrument-defined exposure limits for neutrons and alpha radiation   1:00–00; 2:00–00

1943:   Development of REP and REM units begins   2:00–00

1944:   Provisional tolerance level for plutonium set at 5 micrograms 2:00; 3:00–00

1945:   Plutonium tolerance dose lowered to 1 microgram   3:00–00

1945:   Exposure limit for 100-ton test shot at Trinity set at 0.1 roentgen per day   3:00–00

1945:   Suggested voluntary limit on external exposure for participants in Trinity of 5 roentgens   4:000–000

1945:   Off-site gamma exposure limits for Trinity set at 75 roentgens integrated over 336 hours, 15 roentgens per hour peak   4:000–000

1946:   Exposure limit for Crossroads, 0.1 roentgen per day; up to 10 roentgens in one day or 60 roentgens in two weeks in "special situations" directly approved by Radiological Safety Control 5:00–00

1946:   Ad hoc use of minutes required to receive tolerance dose as safety standard for decontamination crews on Crossroads ships   6:00

1946:   Ad hoc standard for beta radiation based on gamma exposure: permissible beta dose of 0.5 roentgen per day will not be exceeded if gamma exposure does not exceed 0.1 roentgen per day   6:000

1946:   Navy-defined limits for ships exposed at Crossroads: operational clearance, no reading higher than 0.5 roentgen per day combined beta-gamma (0.1 if shielded); final clearance, no reading higher than 0.005 roentgen per day (0.001 shielded)   6:000

NOTES

1. FOUNDATIONS OF MANHATTAN PROJECT RADIATION SAFETY

1. On Röntgen's discovery and its impact, the major source is Otto Glasser, *Wilhelm Conrad Röntgen and the Early History of Röntgen Rays* (Springfield, Ill.: Charles C. Thomas, 1934); but see also George Sarton, "The Discovery of X-Rays," *Isis* 26 (1937):349–364; G. E. M. Jauncey, "The Birth and Early Infancy of X-Rays," *American Journal of Physics* 13 (1945):362–379; E. Ashworth Underwood, "Wilhelm Conrad Röntgen (1845–1923) and the Early Development of Radiology," in Zachary Cope, ed., *Sidelights on the History of Medicine* (London: Butterworth, 1957), 223–241; Alfred Romer, "Accident and Professor Röntgen," *American Journal of Physics* 27 (1959):275–277; and G. L'E. Turner, "Röntgen," in Charles Coulston Gillispie, ed., *Dictionary of Scientific Biography*, vol. 11 (New York: Charles Scribner's Sons, 1975), 529–531.

2. Daniel Paul Serwer, "The Rise of Radiation Protection: Science, Medicine and Technology in Society, 1896–1935," Ph.D. diss., History of Science, Princeton University, 1977, 41. This is by far the best study of its subject I have seen.

3. Ronald L. Kathren, "Early X-Ray Protection in the United States," *Health Physics* 8 (1962):503–511; Kathren, "William H. Rollins (1852–1929): X-Ray Protection Pioneer," *Journal of the History of Medicine and Allied Sciences* 19 (1964):287–294; Kathren, "Historical Development of Radiation Measurement and Protection," in Allen B. Brodsky, ed., *CRC Handbook of Radiation Measurement and Protection*, Section A, vol. 1: *Physical Science and Engineering Data* (West Palm Beach, Fla.: CRC Press, 1978), 13–52; S. J. Vacirca, "Radiation Injuries before 1925," *Radiologic Technology* 39 (1968):347–352; Ruth and Edward Brecher, *The Rays: A History of Radiology in the United States and Canada* (Baltimore: Williams & Wilkins, 1969), 81–90, 161–174.

4. As quoted in Stewart C. Bushong, "The Development of Current Radiation Protection Practices in Diagnostic Radiology," *CRC Critical Reviews in Radiological Sciences* 2 (1971):343.

5. W. D. Coolidge and E. E. Charlton, "Roentgen-Ray Tubes," *Radiology* 45 (1945):449–450; Otto Glasser, "Technical Development of Radiology," *American Journal of Roentgenology* 75 (1956):7–8, 10; Brecher and Brecher, *The Rays*, 44–58 (n. 3).

6. Edith H. Quimby, "The History of Dosimetry in Roentgen Therapy," *American Journal of Roentgenology* 54 (1945):690.

7. Serwer, "Rise of Radiation Protection," 36–37, 43 (n. 2); Quimby, "History of Dosimetry," 688 (n. 6); Herbert M. Parker and William C. Roesch, "Units, Radiation: Historical Development," in George L. Clark, ed., *The Encyclopedia of X-Rays and Gamma Rays* (New York: Reinhold, 1963), 1102; Parker, "Some Background Information on the Development of Dose Units," Report no. CRUSP-1, Commission on Radiologic Units, Standards, and Protection, American College of Radiology, Nov. 1955, 3–9.

8. Mary André Chorzempa, "Ionizing Radiation and Its Chemical Effects: A Historical Study of Chemical Dosimetry (1902–1962)," Ph.D. diss., History of Science, Oregon State University, 1971, chap. 2; Serwer, "Rise of Radiation Protection," chap. 3 (n. 2); Quimby, "History of Dosimetry," 688–692 (n. 6); Gioacchino Failla, "Ionization Measurements," *American Journal of Roentgenology* 10 (1923):48–56 (repr. in *Health Physics* 38 [1980]:889–897).

9. Coolidge and Charlton, "Roentgen-Ray Tubes," 451–452 (n. 5); Glasser, "Technical Development of Radiology," 8, 10 (n. 5); Kathren, "Historical Development," 24–25 (n. 3); Bushong, "Development of Current Radiation Protection Practices," 349–350 (n. 4); Brecher and Brecher, *The Rays,* 191–210 (n. 3); Serwer, "Rise of Radiation Protection," 119–123 (n. 2); Sidney Russ et al., "The Injurious Effects Caused by X-Rays," *Journal of the Roentgen Society* 12 (1916):38–56.

10. Serwer, "Rise of Radiation Protection," 126–131 (n. 2); Lawrence Reynolds, "The History of the Use of the Roentgen Ray in Warfare," *American Journal of Roentgenology* 54 (1945):649–662; Lauriston S. Taylor, *Organization for Radiation Protection: The Operations of the ICRP and NCRP, 1928–1974,* Department of Energy report DOE/TIC-10124 (Washington, 1979), 1-001 to 2-003.

11. Serwer, "Rise of Radiation Protection," 150–172 (n. 2).

12. Sidney Russ, "A Personal Retrospect," *British Journal of Radiology* 26 (1953):554.

13. Serwer, "Rise of Radiation Protection," 172–180 (n. 2); Bushong, "Development of Current Radiation Protection Practices," 345 (n. 4); Vacirca, "Radiation Injuries," 351 (n. 3).

14. F. G. Spear, "The British X-Ray and Radium Protection Committee," *British Journal of Radiology* 26 (1953):553.

15. Taylor, *Organization for Radiation Protection,* 2-003 to 2-010 (n. 10); Serwer, "Rise of Radiation Protection," 175–181 (n. 2).

16. Hermann Wintz, assisted by Walther Rump, *Protective Measures against Danger Resulting from the Use of Radium, Roentgen and Ultra-Violet Rays,* League of Nations Health Organization Document C.H. 1054 (Geneva, 1931), 7.

17. Serwer, "Rise of Radiation Protection," 42, 179–180 (n. 2); Lauriston S. Taylor, *Radiation Protection Standards* (Cleveland: CRC Press, 1971), 12–13.

18. Arthur Mutscheller, "Physical Standards of Protection against Roentgen-Ray Dangers," *American Journal of Radiology* 13 (1925):65.

19. E. R. N. Grigg, *The Trail of the Invisible Light: From X-Strahlen to Radio(bio)logy* (Springfield, Ill.: Charles C. Thomas, 1965), 137, 851–852; Serwer,

"Rise of Radiation Protection," 226–227 (n. 2); Taylor, *Organization for Radiation Protection*, 3-011 (n. 10); Taylor, "The Development of Radiation Protection Standards (1925–1940)," *Health Physics* 41 (1981):228–230.

20. Mutscheller, "Physical Standards of Protection," 65 (n. 18).

21. Ibid., 66–67.

22. Ibid., 67.

23. Ibid., Mutscheller's emphasis.

24. Ibid., 68–70.

25. Lauriston S. Taylor, "History of the International Commission on Radiological Units and Measurements (ICRU)," *Health Physics* 1 (1958):306–307; Taylor, *Organization for Radiation Protection*, 3-010, 3-011 (n. 10); Serwer, "Rise of Radiation Protection," 215–224 (n. 2).

26. Taylor, *Organization for Radiation Protection*, 3-012 to 3-024 (n. 10); Taylor, *Radiation Protection Standards*, 13–16 (n. 17); Taylor, "History of the International Commission on Radiological Protection (ICRP)," *Health Physics* 1 (1958):97–104; Taylor, "Reminiscences about the Early Days of Organized Radiation Protection," in Ronald L. Kathren and Paul L. Ziemer, eds., *Health Physics: A Backward Glance. Thirteen Original Papers on the History of Radiation Protection* (New York: Pergamon Press, 1980), 112–113. See also Serwer, "Rise of Radiation Protection," 224–232 (n. 2).

27. "Recommendations Adopted by the Second International Congress of Radiology," July 1928, repr. in Taylor, *Organization for Radiation Protection*, 3-022 (n. 10).

28. Taylor, "Reminiscences," 114–115 (n. 26); Taylor, *Organization for Radiation Protection*, 4-001 to 4-003 (n. 10); Taylor, "Brief History of the National Committee on Radiation Protection and Measurements (NCRP) Covering the Period 1929–1946," *Health Physics* 1 (1958):3–4.

29. Taylor, *Radiation Protection Standards*, 17–18 (n. 17).

30. Taylor, "Reminiscences," 114 (n. 26); Taylor, *Organization for Radiation Protection*, 4-003, 4-004 (n. 10); *X-Ray Protection* (NBS Handbook 15, 16 May 1931), repr. in ibid., App. J.; Philip M. Boffey, "Radiation Standards: Are the Right People Making Decisions?" *Science* 171 (1971):780, 782.

31. Taylor, *Radiation Protection Standards*, 18 (n. 17).

32. Serwer, "Rise of Radiation Protection," 230–231, 233–234 (n. 2); Taylor, *Radiation Protection Standards*, 13–14 (n. 17); Taylor, *Organization for Radiation Protection*, 4-016 to 4-020 (n. 10); Taylor, "History of the ICRP," 98 (n. 26).

33. *Radium Protection for Amounts up to 300 Milligrams* (NBS Handbook 18, 17 May 1934), repr. in Taylor, *Organization for Radiation Protection*, 4-005 to 4-008, at 4-007 (n. 10).

34. Taylor, *Organization for Radiation Protection*, 4-010 to 4-011, 4-012 to 4-015 (n. 10); Taylor, "Reminiscences," 116–117 (n. 26).

35. The ICRP statement is quoted in Taylor, *Organization for Radiation Protection*, 4-021 (n. 10).

36. Taylor, *Radiation Protection Standards*, 15 (n. 17).

37. Taylor, "Development of Radiation Protection Standards," 231 (n. 19); Serwer, "Rise of Radiation Protection," 233–234 (n. 2). In *Organization for Ra-*

diation Protection, 4-021, and "Technical Accuracy in Historical Writing," *Health Physics* 40 (1981):596–597, Taylor corrects his own earlier account of the widely repeated explanation for the discrepancy between American and international figures; both groups used the same data and merely rounded them differently.

38. As quoted in Taylor, *Organization for Radiation Protection,* 4-021 (n. 10).

39. For a good overall account see Alfred Romer, *The Restless Atom* (Garden City, N.Y.: Anchor Books, 1960). On early events, see also Lawrence Badash, "Chance Favors the Prepared Mind: Henri Becquerel and the Discovery of Radioactivity," *Archives Internationales d'Histoire des Sciences* 70 (1965):55–66; Badash, "Radioactivity before the Curies," *American Journal of Physics* 33 (1965):128–135; and Romer, *The Discovery of Radioactivity and Transmutation* (New York: Dover, 1964). For later developments, see also Thaddeus J. Trenn, "Rutherford on the Alpha-Beta-Gamma Classification of Radioactive Rays," *Isis* 67 (1976):61–75; Romer, *Radioactivity and the Discovery of Isotopes* (New York: Dover, 1970); and Badash, *Radioactivity in America: Growth and Decay of a Science* (Baltimore: Johns Hopkins University Press, 1979).

40. James T. Case, "The Early History of Radium Therapy and the American Radium Society," *American Journal of Roentgenology* 82 (1959):578–579; Edith H. Quimby, "The Background of Radium Therapy in the United States, 1906–1956," *American Journal of Roentgenology* 75 (1956):444; Quimby, "Radium Protection," *Journal of Applied Physics* 10 (1939):604–608. Badash, *Radioactivity in America,* includes excellent summaries of "Radioactivity in Medicine," 125–134, and "The Radium Business," 135–151 (n. 39). See also Serwer, "Rise of Radiation Protection," 9–10 (n. 2); Taylor, *Radiation Protection Standards,* 16, 18 (n. 17); Edward R. Landa, "The First Nuclear Industry," *Scientific American* 247 (Nov. 1982):180ff.

41. Vacirca, "Radiation Injuries," 349–351 (n. 3); Taylor, "History of the ICRU," 309–310 (n. 25); Serwer, "Rise of Radiation Protection," 172, 174–175 (n. 2).

42. Robley D. Evans, "Radium Poisoning: A Review of Present Knowledge," *American Journal of Public Health* 23 (1933):1017–1018.

43. Blum's observation appears as a footnote to the published version of a paper presented in Sept. 1923: Theodore Blum, "Osteomyelitis of the Mandible and Maxilla," *Journal of the American Dental Association* 11 (1924):805.

44. Harrison S. Martland, Philip Conlon, and Joseph P. Knef, "Some Unrecognized Dangers in the Use and Handling of Radioactive Substances, with Especial Reference to the Storage of Insoluble Products of Radium and Mesothorium in the Reticulo-Endothelial System," *Journal of the American Medical Association* 85 (1925):1769–1776; Frederick L. Hoffman, "Radium (Mesothorium) Necrosis," *Journal of the American Medical Association* 85 (1925):961–965; Martland, "Occupational Poisoning in Manufacture of Luminous Watch Dials: General Review of Hazard Caused by Ingestion of Luminous Paint, with Especial Reference to the New Jersey Cases," *Journal of the American Medical Association* 92 (1929):446–473; Evans, "Radium Poisoning," 1018, 1021 (n. 42).

45. Martland, "Occupational Poisoning," 466–467 (n. 44); Evans, "Radium Poisoning," 1019 (n. 42); Evans, "Radium in Man," *Health Physics* 27 (1974):498;

Badash, *Radioactivity in America*, 148–149 (n. 39); Serwer, "Rise of Radiation Protection," 128–129 (n. 2); William D. Sharpe, "The New Jersey Radium Dial Painters: A Classic in Occupational Carcinogenesis," *Bulletin of the History of Medicine* 52 (1979):561, 564; Daniel Lang, "A Most Valuable Accident," *New Yorker*, 2 May 1959, 49ff., repr. in Lang, *From Hiroshima to the Moon* (New York: Dell, 1961), 423–424.

46. Wilfrid B. Mann and S. B. Garfinkel, *Radioactivity and Its Measurement* (Princeton, N.J.: D. Van Nostrand, 1966), chap. 2; and Samuel Glasstone, ed., *Sourcebook on Atomic Energy* (New York: D. Van Nostrand, 1950), chap. 5.

47. Evans, "Radium Poisoning," 1019 (n. 42).

48. Hoffman, "Radium Necrosis," 963–964 (n. 44); Martland, "Occupational Poisoning," 468 (n. 44); Sharpe, "Dial Painters," 561–563 (n. 45).

49. William B. Castle, Katherine R. Drinker, and Cecil K. Drinker, "Necrosis of the Jaw in Workers Employed in Applying a Luminous Paint Containing Radium," *Journal of Industrial Hygiene* 7 (1925):317–382.

50. Hoffman, "Radium Necrosis," 961–962 (n. 44); Sharpe, "Dial Painters," 563–564; Roger J. Cloutier, "Florence Kelley and the Radium Dial Painters," *Health Physics* 39 (1980):712–713.

51. Martland, "Occupational Poisoning," 466 (n. 44).

52. Martland, Conlon, and Knef, "Some Unrecognized Dangers," 1774 (n. 44).

53. Martland, "Occupational Poisoning," 469 (n. 44). See also Lang, "A Most Valuable Accident," 424–425 (n. 45).

54. Harrison S. Martland, "The Occurrence of Malignancy in Radioactive Persons," *American Journal of Cancer* 15 (1931):2435–2516.

55. Evans, "Radium Poisoning," 1019 (n. 42).

56. Sharpe, "Dial Painters," 567–569 (n. 45); Lang, "A Most Valuable Accident," 430–445 (n. 45); Cloutier, "Florence Kelley," 713–715 (n. 50); Martland, "Occupational Poisoning," 471–473 (n. 44).

57. Anthony P. Polednak, Andrew F. Stehney, and R. E. Rowland, "Mortality among Women First Employed before 1930 in the U.S. Radium Dial-Painting Industry," *Journal of Epidemiology* 107 (1978):179–195; Polednak, "Bone Cancer among Female Radium Dial Workers: Latency Periods and Incidence Rates by Time after Exposure," *Journal of the National Cancer Institute* 60 (1978):77–82.

58. Martland, "Occupational Poisoning," 470–473 (n. 44); G. S. Reitter and Martland, "Leucopenic Anemia of the Regenerative Type Due to Exposure to Radium and Mesothorium," *American Journal of Roentgenology* 16 (1926):161–167; Evans, "Radium in Man," 498–499 (n. 45); William D. Sharpe, "Chronic Radium Intoxication: Clinical and Autopsy Findings in Long-Term New Jersey Survivors," *Environmental Research* 8 (1974):243–383; R. E. Rowland, A. F. Stehney, A. M. Brues, M. S. Littman, A. T. Keane, B. C. Patten, and M. M. Shanahan, "Current Status of the Study of ^{226}Ra and ^{228}Ra in Humans at the Center for Human Radiobiology," *Health Physics* 35 (1978):159–160.

59. Evans, "Radium in Man," 499 (n. 45). See also Fritz Schales, "Brief History of ^{224}Ra Usage in Radiotherapy and Radiobiology," *Health Physics* 35 (1978):25–32.

60. Evans, "Radium Poisoning," 1018–1019 (n. 42); Evans, "Origin of Standards for Internal Emitters," in Kathren and Ziemer, eds., *Health Physics*, 142 (n. 26); Evans, "Inception of Standards for Internal Emitters, Radon and Radium," *Health Physics* 41 (1981):437; Kathren, "Historical Development," 35–36 (n. 3).

61. Evans, "Origin of Standards," 141–143, 146 (n. 60); Evans, "Inception of Standards," 437–438 (n. 60); interview with Robley D. Evans, Scottsdale, Ariz., 8 Nov. 1978 (all interviews, unless otherwise noted, were conducted by the author).

62. Evans, "Origin of Standards," 143–147 (n. 60); Evans, "Inception of Standards," 438–441 (n. 60); Evans, "Protection of Radium Dial Workers and Radiologists from Injury by Radium," *Journal of Industrial Hygiene and Toxicology* 25 (1943):253–269; Wright H. Langham and John W. Healy, "Maximum Permissible Body Burdens and Concentrations of Plutonium: Biological Basis and History of Development," in Harold C. Hodge, J. Newell Stannard, and J. B. Hursh, eds., *Uranium, Plutonium, Transplutonic Elements*, Handbook of Experimental Pharmacology, vol. 36 (New York: Springer-Verlag, 1973), 571.

63. Evans, "Origin of Standards," 147–148 (n. 60); Evans, "Inception of Standards," 441, 443 (n. 60).

64. *Safe Handling of Radioactive Luminous Compound* (NBS Handbook 27, 2 May 1941); Taylor, *Organization for Radiation Protection*, 5-022 to 5-025 (n. 10).

65. Robley D. Evans and Clark Goodman, "Determination of the Thoron Content of Air and Its Bearing on Lung Cancer Hazards in Industry," *Journal of Industrial Hygiene and Toxicology* 22 (1940):89–99 (repr. in *Health Physics* 38 [1980]:919–928). See also Evans, "Origin of Standards," 148 (n. 60); Taylor, *Organization for Radiation Protection*, 5-022 (n. 10).

66. Herbert J. Muller, "Artificial Transmutation of the Gene," *Science* 66 (1927):84–87; Gioacchino Failla, "Biological Effects of Ionizing Radiation," *Journal of Applied Physics* 12 (1941):279–295. The failure of current standards to deal with genetic damage is the central thesis of Paul S. Henshaw, "Biological Significance of the Tolerance Dose in X-Ray and Radium Protection," *Journal of the National Cancer Institute* 1 (1941):789–805. See also Elof Axel Carlson, *Genes, Radiation, and Society: The Life and Work of H. J. Muller* (Ithaca, N.Y.: Cornell University Press, 1981); Daniel S. Grosch and Larry E. Hopwood, *Biological Effects of Radiations* (2d ed.; New York: Academic Press, 1979), 115–139.

67. Advisory Committee on X-Ray and Radium Protection, "Report," 3 Dec. 1940, in Taylor, *Organization for Radiation Protection*, 5-011 to 5-013 (n. 10); "Notes on Meeting of Advisory Protection Committee, Cleveland," 4 Dec. 1940, ibid., 5-013 to 5-015.

68. Failla letter, "Subject: Remarks by Dr. Failla," 16 June 1941, in Taylor, *Organization for Radiation Protection*, 5-016 (n. 10).

69. Advisory Committee on X-Ray and Radium Protection, Minutes of meeting, Cincinnati, 25 Sept. 1941, in Taylor, *Organization for Radiation Protection*, 5-018 to 5-021 (n. 10).

70. Lauriston S. Taylor, "X-Ray Protection," *Journal of the American Medical Association* 116 (1941):139.

71. See, e.g., Henshaw, "Biological Significance," 792 (n. 66), citing Taylor, "X-Ray Protection," 136–140 (n. 70).

72. Advisory Committee minutes, 25 Sept. 1941, 5-020 (n. 69).

73. On the interwar history of physics in the context of the American physics community, see Daniel J. Kevles, *The Physicists: The History of a Scientific Community in Modern America* (New York: Alfred A. Knopf, 1977), chaps. 15–18, with a very full bibliographical essay. Three studies cover the wartime programs: David Irving, *The German Atomic Bomb: The History of Nuclear Research in Nazi Germany* (New York: Simon & Schuster, 1967); Margaret Gowing, *Britain and Atomic Energy, 1939–1945* (London: Macmillan; New York: St. Martin's Press, 1964); Richard G. Hewlett and Oscar E. Anderson, Jr., *A History of the United States Atomic Energy Commision*, vol. 1: *The New World, 1939/1946* (University Park: Pennsylvania State University Press, 1962).

74. Hewlett and Anderson, *New World*, chap. 3 (n. 73). See also Arthur H. Compton, *Atomic Quest: A Personal Narrative* (New York: Oxford University Press, 1956); Laura Fermi, *Atoms in the Family: My Life with Enrico Fermi* (Chicago: University of Chicago Press, 1954).

75. Robert S. Stone, "General Introduction to Reports on Medicine, Health Physics, and Biology," in Stone, ed., *Industrial Medicine on the Plutonium Project: Survey and Collected Papers*, National Nuclear Energy Series, Div. IV, vol. 20 (New York: McGraw-Hill, 1951), 1–2; Hewlett and Anderson, *New World*, chap. 3 (n. 73).

76. Merril Eisenbud, *Environmental Radioactivity* (2d ed.; New York: Academic Press, 1973), 41–42. Cf. Failla, "Biological Effects," 279–295 (n. 66); Henshaw, "Biological Significance," 789–805 (n. 66).

77. This phrase appears more than once in NBS Handbook 27 (n. 64); see Taylor, *Organization for Radiation Protection*, 5-022 (n. 10).

78. Dean B. Cowie and Leonard A. Scheele, "A Survey of Radiation Protection in Hospitals," *Journal of the National Cancer Institute* 1 (1941):767–787 (repr. in *Health Physics* 38 [1980]:929–947), discuss some of the uncertainties and also tend to show that existing hazards resulted chiefly from failures to adhere to standards. See also Failla, "Biological Effects," 281–284 (n. 66); Henshaw, "Biological Significance" (n. 66).

79. Interview with Leon O. Jacobson, Chicago, 14 June 1979.

80. Stone, "General Introduction," 15 (n. 75); Leon O. Jacobson and Edna K. Marks, "The Hematological Effects of Ionizing Radiations in the Tolerance Range," *Radiology* 49 (1947):286–298; Jacobson and Marks, "Clinical Laboratory Examination of Plutonium Project Personnel," in Stone, ed., *Industrial Medicine*, 113–139 (n. 75); Jacobson, Marks, and Egon Lorenz, "Hematological Effects of Ionizing Radiation," ibid., 140–196; G. A. Sacher, Jacobson, Marks and S. L. Tylor, "Biometric Investigations of Blood Constituents and Characteristics in a Population of Project Workers," ibid., 425–455.

81. Compton, *Atomic Quest*, 177 (n. 74).

82. Interview with Kenneth S. Cole, San Diego, 1 Nov. 1979. See also Cole to author, 27 March 1982 (letter in author's possession); Stone, "General Introduction," 15 (n. 75).

83. John H. Lawrence to C. D. Shane, 10 Jan. 1944 (E. O. Lawrence Collection, Bancroft Library, University of California, Berkeley, carton 2, folder 29); Stafford L. Warren, "The Role of Radiology in the Development of the Atomic Bomb," in Kenneth D. A. Allen, ed., *Radiology in World War II*, Medical Department in World War II: Clinical Series (Washington: Office of the Surgeon General, U.S. Army, 1966), 858–864; Brecher and Brecher, *The Rays*, 348–350, 367–370, 385–387, 403–404 (n. 3).

84. Compton, *Atomic Quest*, 177 (n. 74).

85. Stone, "General Introduction," 15 (n. 75).

86. Ibid.; Jacobson interview (n. 79).

87. Interview with Raymond D. Finkle, Los Angeles, 17 April 1980.

88. Interview with Ernest O. Wollan, Edina, Minn., 4 Sept. 1979; Stone, "General Introduction," 3 (n. 75); Hewlett and Anderson, *New World*, 206 (n. 73). For an example of this early work, see Met Lab report CH-137, "Protection against Radiations; Protection against γ Rays," by E. P. Wigner, n.d., but ca. 20 June 1942. Copies of most Met Lab reports are on file at the Records Center, Argonne National Laboratory, Argonne, Ill.; they are arranged by report number, i.e., chronologically, as reports were numbered serially from the project's start. All cited reports have been declassified.

89. Kevles, *The Physicists*, chap. 21 and passim (n. 73); Carroll W. Pursell, Jr., "Science Agencies in World War II: The OSRD and Its Challengers," in Nathan Reingold, ed., *The Sciences in the American Context: New Perspectives* (Washington: Smithsonian Institution Press, 1979), 359–378; A. Hunter Dupree, "The Great *Instauration* of 1940: The Organization of Scientific Research for War," in Gerald Holton, ed., *The Twentieth-Century Sciences: Studies in the Biography of Ideas* (New York: W. W. Norton, 1972), 443–467.

90. Cole interview (n. 82); interviews with James J. Nickson, Memphis, Tenn., 12 July 1979; Herbert M. Parker, Richland, Wash., 5 Oct. 1979; Richard Abrams, Pittsburgh, 9 May 1980; and Waldo E. Cohn, Oak Ridge, Tenn., 18 July 1979.

91. Interviews with Eric L. Simmons, Chicago, 14 May 1980; Charles W. Hagen, Jr., Bloomington, Ind., 20 May 1980; Hagen to author, 18 Feb. 1981.

92. Interview with C. Ladd Prosser, Urbana, Ill., 31 Aug. 1979.

93. Compton, *Atomic Quest*, 178–179 (n. 74).

94. Met Lab report CH-376, "Health and Safety," 26 Nov. 1942.

95. Stone, "General Introduction," 5 (n. 75).

2. ROLE OF THE CHICAGO HEALTH DIVISION

1. Met Lab report CH-237, "Health Division Report for Month Ending August 15, 1942," n.d.

2. Met Lab report CH-259, "Health Division Report for Month Ending September 15, 1942," n.d., report by Simeon T. Cantril and Kenneth S. Cole, 1.

3. Ibid.

4. Robert S. Stone, "Health Protection Activities of the Plutonium Project," *American Philosophical Society Proceedings* 90 (1946):14–15.

5. Cantril and Cole in CH-259, 2–3 (n. 2).

6. Albert E. Tannenbaum, ed., *Toxicology of Uranium: Survey and Collected Papers,* National Nuclear Energy Series, Div. IV, vol. 23 (New York: McGraw-Hill, 1951), especially Tannenbaum and Herbert Silverstone, "Introduction and General Considerations," 3–5, and "Summary of Experimental Studies: Relation to Uranium Poisoning in Man," 45–50; Waldo E. Cohn to author, 3 March 1981; J. Newell Stannard to author, 27 March 1981. The Rochester findings required four fat volumes to describe: Carl Voegtlin and Harold C. Hodge, eds., *Pharmacology and Toxicology of Uranium Compounds, with a Section on the Pharmacology and Toxicology of Fluorine and Hydrogen Fluoride,* National Nuclear Energy Series, Div. VI, vol. 1, parts 1–4 (New York: McGraw-Hill, 1949–1953).

7. Cantril and Cole in CH-259, 2 (n. 2); interviews with John L. Ferry, East Chicago, Ind., 13 May 1980; and James H. Sterner, Irvine, Cal., 16 April 1980. See also W. Daggett Norwood, "Study of Chemical Hazards in Extraction Plant," 24 April 1943 (copy in Met Lab Reading File, Argonne National Laboratory [hereafter ANL Reading File]); Met Lab report CH-732, "Site X Operating Manual: Health Hazards," May 1943, 25–33.

8. Cantril and Cole in CH-259, 2 (n. 2).

9. Leon O. Jacobson and Edna K. Marks, "Clinical Laboratory Examination of Plutonium Project Personnel," in Robert S. Stone, ed., *Industrial Medicine on the Plutonium Project: Survey and Collected Papers,* National Nuclear Energy Series, Div. IV, vol. 20 (New York: McGraw-Hill, 1951), 113–139.

10. Ernest O. Wollan in CH-259, 1 (n. 2).

11. Ralph E. Lapp, "Survey of Nucleonics Instrumentation Industry," *Nucleonics* 4 (May 1949):101; Ronald L. Kathren, "Before Transistors, IC's, and All Those Other Good Things: The First Fifty Years of Radiation Monitoring Instrumentation," in Kathren and Paul L. Ziemer, eds., *Health Physics: A Backward Glance. Thirteen Original Papers on the History of Radiation Protection* (New York: Pergamon Press, 1980), 79; Richard D. Terry, "Historical Development of Commercial Health Physics Instrumentation," ibid., 160; Herbert M. Parker to author, 10 March 1981.

12. Stone to Emilio Segrè, 27 March 1943; R. B. Smith and Herbert M. Parker to M. D. Whitaker, "Pocket and Film Meters for Use at Clinton Laboratories," 24 Aug. 1943 (both ANL Reading File); Parker to W. W. Watson, 1 Feb. 1944 (in Chronological File of Miscellaneous Documents [Historical Interest], Argonne National Laboratory; cited hereafter as ANL Chron File); interviews with John E. Rose, Mountain Home, Ark., 11 July 1979, and Howard C. Eberline, Edmond, Okla., 28–29 Sept. 1978; John W. Healy to author, 19 May 1981; Adrian H. Dahl to author, 12 March 1981. See also Stone, "Health Protection," 14 (n. 4); James J. Nickson, "Protective Measures for Personnel," in Stone, ed., *Industrial Medicine,* 110 (n. 9); Karl Z. Morgan, "Instrumentation in the Field of Health Physics," *Proceedings of the I.R.E.* 37 (Jan. 1949):75.

13. Wollan in CH-259, 1 (n. 10); Arthur W. Fuchs, "Evolution of Roentgen Film," *American Journal of Roentgenology* 75 (1956):40–44; Mary André Chorzempa, "Ionizing Radiation and Its Chemical Effects: A Historical Study of Chemical Dosimetry (1902–1962)," Ph.D. diss., History of Science, Oregon State University, 1971, 80–83.

14. Edith H. Quimby, "Radium Protection," *Journal of Applied Physics* 10 (1939):608.

15. Met Lab report CH-1553, "Photographic Film as a Pocket Radiation Dosimeter," by L. A. Pardue, N. Goldstein, and Ernest O. Wollan, 8 April 1944. See also Klaus Becker, *Photographic Film Dosimetry: Principles and Methods of Quantitative Measurement of Radiation by Photographic Means*, trans. K. S. Ankersmith (London and New York: Focal Press, 1966), 13–15.

16. Wollan in CH-259, 1 (n. 10).

17. CH-1553, 1–4 (n. 15); Radiation Monitoring and Instrument Coordinating Committee to Stone, "Plans for Monitoring of Radiation for Health Protection at the Clinton Laboratories," 15 May 1943 (ANL Reading File); Smith and Parker to Whitaker, 24 Aug. (n. 12).

18. Wollan in CH-259, 1 (n. 10).

19. Stone, "Health Protection," 16–17 (n. 4). See also Stone to R. R. Newell, 7 April 1943; Stone to Paul C. Aebersold, 15 April 1943; Stone to Arthur Holley Compton, 15 May 1943; Stone to Paul C. Henshaw, 25 May 1943 (ANL Reading File).

20. Stone to Joseph G. Hamilton, 26 May 1944 (E. O. Lawrence Collection, Bancroft Library, University of California, Berkeley, carton 28, folder 40; cited hereafter as Lawrence Coll. c/f), exemplifies Stone's repeated strictures about intake. See also Stone, "Health Protection," 17 (n. 4); Stone, "General Introduction to Reports on Medicine, Health Physics, and Biology," in Stone, ed., *Industrial Medicine*, 9 (n. 9).

21. Simeon T. Cantril, "Biological Bases for Maximum Permissible Exposures," in Stone, ed., *Industrial Medicine*, 48 (n. 9).

22. Stone, "General Introduction," 3–4 (n. 20).

23. Samuel K. Allison to Miles Leverett, "Permissible Dosage of Penetrating Radiation," 1 April 1943 (ANL Reading File). The change is neatly illustrated in the title of Cantril's paper (n. 21); in Met Lab report CH-2812, written jointly by Cantril and Parker in 1943 (although issued 5 Jan. 1945), it was entitled "The Tolerance Dose."

24. Wollan in CH-259, 1 (n. 10); Wollan, "Problems of the Physics Section," in Stone et al., "Report on Experimental Program of Health Division," April 1943 (ANL Reading File); William P. Jesse, "The Role of Instruments in the Atomic Bomb Project," *Chemical Engineering News* 24 (1946): 2906–2909; Lapp, "Nucleonics Instrumentation Industry" (n. 11); interviews with Carl C. Gamertsfelder, Knoxville, Tenn., 13 July 1979, and Donald L. Collins, Glendale, Cal., 31 Oct. 1979.

25. Jesse, "Role of Instruments," 2906 (n. 24).

26. In addition to the works cited in note 11, above, see also Jesse, "Role of Instruments," 2906 (n. 24); F. R. Shonka and H. L. Wyckoff, "Measurement of

Nuclear Radiations," in Samuel Glasstone, ed., *The Effects of Atomic Weapons* (Washington: Government Printing Office, 1950), 291–311; Glasstone, ed., *Sourcebook on Atomic Energy* (New York: D. Van Nostrand, 1950), chap. 6; Edwin A. Bemis, Jr., "Survey Instruments and Pocket Dosimeters," in Gerald J. Hine and Gordon L. Brownell, eds., *Radiation Dosimetry* (New York: Academic Press, 1956), 454–503; Wilfrid B. Mann and S. B. Garfinkel, *Radioactivity and Its Measurement* (Princeton, N. J.: D. Van Nostrand, 1966), chap. 6.

27. In addition to the works cited in notes 11 and 26, above, see also Nicholas Anton, "Radiation Counter Tubes and Their Operation," *Electronic Industries and Electronic Instrumentation* 2 (Feb. 1948):4–7; H. Greinacher, "The Evolution of Particle Counters," *Endeavour* 13 (Oct. 1954):190–197.

28. Jesse, "Role of Instruments," 2908 (n. 24).

29. Ibid.; see also the works cited in note 26, above.

30. Stone to Compton, 10 April 1943 (ANL Reading File).

31. Stone to Aebersold, 15 April (n. 19); L. B. Arnold, Jr., to T. R. Hogness, "Tolerance Levels for Radiation," 25 Aug. 1944, attachment 4, Arnold to E. D. Eastman et al., "Metallurgical Laboratory Procedures for Controlling Product [Plutonium] Hazards," 11 Oct. 1944 (in Mail and Records Center, Los Alamos Scientific Laboratory, Los Alamos, N.M.); Wollan in Met Lab report CH-632, "Health Division Program," 10 May 1943, 13–14.

32. Parker, "Some Background Information on the Development of Dose Units," Report no. CRUSP-1, Commission on Radiologic Units, Standards, and Protection, American College of Radiology, Nov. 1955, 34.

33. Parker, "Some Background Information," 32–35 (n. 32); CH-2812, 3–13 (n. 23); Parker and William C. Roesch, "Units, Radiation: Historical Development," in George L. Clark, ed., *The Encyclopedia of X-Rays and Gamma Rays* (New York: Reinhold, 1963), 1105; Parker, "Protection Programs of the Plutonium Project," *Health Physics* 41 (1981):572.

34. Wright H. Langham and John W. Healy, "Maximum Permissible Body Burdens and Concentrations of Plutonium: Biological Basis and History of Development," in Harold C. Hodge, J. Newell Stannard, and J. B. Hursh, eds., *Uranium, Plutonium, Transplutonic Elements,* Handbook of Experimental Pharmacology, vol. 36 (New York: Springer-Verlag, 1973), 574–575.

35. Parker, "Tentative Dose Units for Mixed Radiations," presented at annual meeting of the Radiological Society of North America, San Francisco, 5–10 Dec. 1948; Parker, "Health-Physics, Instrumentation, and Radiation Protection," *Advances in Biological and Medical Physics* 1 (1948):242–243. The San Francisco paper was later published in *Radiology* 54 (1950):257–261; and both papers have been reprinted in the special 25th anniversary issue of *Health Physics* 38 (June 1980). See also Glasstone, ed., *Sourcebook,* 504–506 (n. 26); Daniel S. Grosch and Larry E. Hopwood, *Biological Effects of Radiations* (2d ed.; New York: Academic Press, 1979), 13.

36. Cantril in CH-632, 608 (n. 31); A. H. Compton to Arthur V. Peterson, "Contracts Sponsored by Metallurgical Laboratory," 7 April 1943 (ANL Reading File); L. F. Craver, "Tolerance to Whole-Body Irradiation of Patients with Advanced Cancer," in Stone, ed., *Industrial Medicine,* 485–498 (n. 9).

37. Cantril, "Clinical Medicine and Research," 2, in "Report on Experimental Program" (n. 24).

38. Interview with Lee Wattenberg, Minneapolis, 26 June 1979.

39. Cantril and Cole in CH-259 (n. 2); Stone in "Report on Experimental Program," 1–2 (n. 24); Compton to Peterson, 7 April (n. 36). See also Egon Lorenz and Walter E. Heston, "Effects of Long-Continued Total-Body Gamma Irradiation on Mice, Guinea Pigs, and Rabbits, 1. Preliminary Experiments," in Raymond E. Zirkle, ed., *Biological Effects of External X and Gamma Radiation,* National Nuclear Energy Series, Div. IV, vol. 22B (New York: McGraw-Hill, 1954), 1–11.

40. Hamilton, "A Projected Study of the Metabolism of Fission Products from Uranium," in CH-259, 1–3 (n. 2).

41. Hamilton to Stone, 20 Feb. 1943 (Lawrence Coll. 28/40); Met Lab Organization Chart, 23 Jan. 1943 (ANL Chron File); Compton to Peterson, 7 April 1943 (n. 36). See also Hamilton, "A Review of the Research upon the Metabolism of Long-Life Fission Products," 8 June 1943 (Lawrence Coll. 28/40); Hamilton to Samuel K. Allison, "Plans for Future Biological Research," 11 Sept. 1945 (Lawrence Coll. 5/10); and Hamilton, "The Metabolism of the Fission Products and the Heaviest Elements," *Radiology* 49 (1947):325–343.

42. Stone in CH-632, 1 (n. 31).

43. Stone, "General Introduction," 4 (n. 20).

44. Cantril and Cole in CH-259, 3–4 (n. 2).

45. Wollan in CH-259, 1–2 (n. 10); Cantril, "Biological Bases," 65 (n. 21); interviews with Richard Abrams, Pittsburgh, 9 May 1980; Eric L. Simmons, Chicago, 14 May 1980; and J. Garrott Allen, Stanford, Cal., 15 April 1980.

46. Stone in CH-632, 1 (n. 31).

47. Interview with Samuel Schwartz, Minneapolis, 25 June 1979.

48. For a convenient overview of some of the main lines of research pursued during the war, see the symposium on "The Plutonium Project," presented at the annual meeting of the Radiological Society of North America, Chicago, 1–6 Dec. 1946 (published ·in *Radiology* 49 [1947]:269–365).

49. Gamertsfelder interview (n. 24).

50. Stone, "General Introduction," 5 (n. 20).

51. A. H. Compton to James B. Conant, "Absence of Public Hazards of Nuclear Physics Work at University of Chicago and at Argonne Forest," 11 Jan. 1943 (ANL Chron File).

52. Richard G. Hewlett and Oscar E. Anderson, Jr., *A History of the United States Atomic Energy Commission,* vol. 1: *The New World, 1939/1946* (University Park: Pennsylvania State University Press, 1962), 115–119, 185–189, 198. The Clinton Engineer Works was also the site for the major U 235 separation plants: Y-12 (electromagnetic separation); K-25 (gaseous diffusion); and S-50 (liquid-thermal-diffusion). See also Charles W. Johnson and Charles O. Jackson, *City behind a Fence: Oak Ridge, Tennessee, 1942–1946* (Knoxville: University of Tennessee Press, 1981); Paul Loeb, *Nuclear Culture: Living and Working in the World's Largest Atomic Complex* (New York: Coward, McCann & Geoghegan, 1982), pt. 1.

53. Hewlett and Anderson, *New World*, 190–192, 207, 210 (n. 52); Arthur Holley Compton, *Atomic Quest: A Personal Narrative* (New York: Oxford University Press, 1956).

54. Hewlett and Anderson, *New World*, 186, 191 (n. 52); Leslie R. Groves, *Now It Can Be Told: The Story of the Manhattan Project* (New York and Evanston: Harper & Row, 1962), 58; Stephane Groueff, *Manhattan Project: The Untold Story of the Making of the Atomic Bomb* (Boston and Toronto: Little, Brown, 1967), chap. 10; Lenore Fine and Jesse A. Remington, *The Corps of Engineers: Construction in the United States*, United States Army in World War II: The Technical Services (Washington: Office of the Chief of Military History, 1972), 666–667; Vance E. Senecal, "Du Pont and Chemical Engineering in the Twentieth Century," in William F. Furter, ed., *History of Chemical Engineering* (Washington: American Chemical Society, 1980), 293.

55. Hewlett and Anderson, *New World*, 115–117, 198–200 (n. 52); Groves, *Now It Can Be Told*, chap. 4 (n. 54); Fine and Remington, *Corps of Engineers*, chap. 20 (n. 54).

56. Stone, "General Introduction," 5 (n. 20); Stafford L. Warren, "The Role of Radiology in the Development of the Atomic Bomb," in Kenneth D. A. Allen, ed., *Radiology in World War II*, Medical Department in World War II: Clinical Series (Washington: Office of the Surgeon General, U.S. Army, 1966), 849; interview with Hymer L. Friedell, Cleveland, 12 June 1979.

57. Henry DeWolf Smyth, *Atomic Energy for Military Purposes: The Official Report on the Development of the Atomic Bomb under the Auspices of the United States Government, 1940–1945* (Princeton: Princeton University Press, 1945), 4.27.

58. Stone, "General Introduction," 4 (n. 20); David Irving, *The German Atomic Bomb: The History of Nuclear Research in Nazi Germany* (New York: Simon & Schuster, 1967), 132–135. See also Arthur V. Peterson, "Peppermint," in Anthony Cave Brown and Charles B. MacDonald, eds., *The Secret History of the Atomic Bomb* (New York: Dial Press/James Wade, 1977), 234. Most of the material in the book, including Peterson's account of radiological safety preparations for D day, comes from detailed histories of all MED projects prepared for Groves after the war.

59. Hewlett and Anderson, *New World*, 37–38 (n. 52).

60. Report by Eugene Wigner and Smyth, Dec. 1941, as quoted in Smyth, *Atomic Energy for Military Purposes*, 4.27 (n. 57).

61. Compton to Vannevar Bush, 22 June 1942 (ANL Chron File).

62. Compton to J. C. Stearns, "Initial Statement of Problems and Duties," 16 July 1942 (ANL Chron File).

63. Darol K. Froman, "Protection Project," in CH-237 (n. 1).

64. Met Lab report CH-259a, "Counter or Defense Measures," by J. C. Stearns, issued as supplement to CH-259 (n. 2).

65. Stearns to Carl Anderson, California Institute of Technology; to L. F. Curtiss, National Bureau of Standards; to H. A. Wilson, Rice Institute, Houston; to Ross Gunn, Naval Research Laboratory, Washington, all dated 15 Sept. 1942 (ANL Chron File).

66. Peterson, "Peppermint," 235 (n. 58); Collins interview (n. 24).

67. Interview with James J. Nickson, Memphis, Tenn., 12 July 1979.

68. Samuel K. Allison to Wollan, 21 May 1943 (ANL Reading File); James B. Conant to Leslie R. Groves, "Preliminary Statement Concerning the Probability of the Use of Radioactive Materials in Warfare," 1 July 1943, as quoted in Irving, *German Atomic Bomb*, 182–183 (n. 58).

69. R. S. Apple to C. M. Cooper, "Availability of Fission Products for Military Purposes," 12 May 1943, with "Supplement," 15 May 1943; Stone to Hamilton, 1 June 1943 (ANL Reading File).

70. Stearns to Peterson, "New Radioactive By-Products," 25 Nov. 1942 (ANL Chron File).

71. Peterson, "Peppermint," 237–238 (n. 58). See also Donald L. Collins, "Pictures from the Past: Journeys into Health Physics in the Manhattan District and Other Diverse Places," in Kathren and Ziemer, eds., *Health Physics*, 40 (n. 11); Groves, *Now It Can Be Told*, 199–206 (n. 54); Lapp, "Nucleonics Instrumentation Industry," 102 (n. 11); Dahl to author (n. 12).

72. Irving, *German Atomic Bomb*, 222–225 and passim (n. 58); Boris T. Pash, *The Alsos Mission* (New York: Charter Books, 1980; repr. of 1969 ed.); S. A. Goudsmit, *Alsos* (New York: Henry Schumann, 1947).

73. Compton to Marshall, "Health Program at Chicago and the Clinton Laboratories," 24 May 1943 (ANL Reading File).

74. Stone, "General Introduction," 5 (n. 20).

75. Compton to Marshall, 24 May (n. 73).

76. Compton to Stone, "Responsibility for Health Program at Site X," 2 April 1943 (ANL Reading File).

77. Stone, "General Introduction," 6 (n. 20).

78. Warren, "Role of Radiology," 848–849 (n. 56); Hewlett and Anderson, *New World*, 148–149 (n. 52); interview with Stafford L. Warren, Los Angeles, 30 Oct. 1979; Friedell interview (n. 56).

79. Nichols to Warren, 10 Aug. 1943, as quoted in Warren, "Role of Radiology," 843 (n. 56).

80. Warren, "Role of Radiology," 841–844 (n. 56); Groves, *Now It Can Be Told*, 421 (n. 54); Fine and Remington, *Corps of Engineers*, 681–682 (n. 54); Warren interview (n. 78); Friedell interview (n. 56).

81. Voegtlin and Hodge, eds., *Uranium Compounds*, especially Andrew H. Dowdy, "University of Rochester Project Foreword," xi–xii; and Hodge, "Historical Foreword," 1–14 (n. 6).

82. Friedell interview (n. 56).

83. Compton, *Atomic Quest*, 179 (n. 53).

84. Stone et al., "Report on Experimental Program" (n. 24); CH-632 (n. 31).

85. Stone to Hamilton et al., "Re: Report to A. H. Compton," 14 Aug. 1945 (Lawrence Coll. 5/10).

86. Hewlett and Anderson, *New World*, 323 (n. 52).

87. Stone to Hamilton et al., 14 Aug. (n. 85).

88. Warren, Memorandum to the files, "Purpose and Limitations of the Biological and Health Physics Research Program," 30 July 1945 (Lawrence Coll. 28/41).

89. Stone, Memo to files, "Col. Stafford L. Warren's Memo Entitled 'Purposes and Limitations of the Biological and Health-Physics Program,'" 17 Aug. 1945 (ANL Chron File).

90. Stone memo to files, 17 Aug. (n. 89).

91. Warren memo to files, 30 July (n. 88).

92. Friedell, Memo to files, "Comment on Tolerance Values for Radium and Product [Plutonium]," 11 May 1945.

93. Friedell interview (n. 56); interviews with Kenneth S. Cole, San Diego, 1 Nov. 1979; and Leon O. Jacobson, Chicago, 14 June 1979.

94. Kenneth Cole to author, 27 March 1982.

95. John E. Wirth, "Medical Services of the Plutonium Project," in Stone, ed., *Industrial Medicine*, 20 (n. 9).

96. "Lecture Schedule—Du Pont Training Program," 19 April 1943 (ANL Reading File).

97. Met Lab report CH-908, "Report of Health Division for Month Ending September 4, 1943," 13–14; Wollan, "Custody of Radium Sources, Radiation Hazard Surveys and Radiation Meters," mimeo, 22 July 1943 (ANL Reading File).

98. Cole to Peterson, 22 April 1943; Curtis, "Biological Measurement to Be Made at Site X," 4 June 1943 (ANL Reading File).

99. Szilard to Vannevar Bush, 14 Jan. 1944, in Spencer R. Weart and Gertrud Weiss Szilard, eds., *Leo Szilard: His Version of the Facts. Selected Recollections and Correspondence* (Cambridge: MIT Press, 1980), 161.

100. Parker to author (n. 11).

101. Stone to M. D. Whitaker, 2 Aug. 1943 (ANL Reading File); Roger Williams to Gen. Groves et al., "Radioactivity Health Hazards—Hanford," 26 June 1944 (ANL Chron File); Hewlett and Anderson, *New World*, 220 (n. 52); interviews with John W. Healy, Los Alamos, 5 June 1980; W. Daggett Norwood, Richland, Wash., 3 Oct. 1979; and Philip A. Fuqua, Richland, Wash., 3 Oct. 1979.

102. Glenn T. Seaborg to Stone, "Physiological Hazards of Working with Plutonium," 5 Jan. 1944, as reproduced in Langham and Healy, "Maximum Permissible Body Burdens," 572–573 (n. 34); the paper by Langham and Healy includes the best available review of the development of plutonium standards during the war at 572–577. See also Hamilton to Stone, 18 Oct. 1944, w/att. "Estimated Tolerance Doses for Product [Plutonium]" (Lawrence Coll. 5/9), Hewlett and Anderson, *New World*, 211–212 (n. 52). The 5-microgram body burden was far from the last word on the subject; see also chap. 4, below.

103. S. T. Cantril, "Industrial Medical Program—Hanford Engineer Works," in Stone, ed., *Industrial Medicine*, 289 (n. 9).

104. Wirth, "Medical Services," 20 (n. 95); Williams to Groves et al., 26 June (n. 101); Stone, "General Introduction," 15 (n. 20).

105. Wirth, "Medical Services," 31 (n. 95).

106. Warren, "Role of Radiology," 846–847, 870–875 (n. 56).

107. Wirth, "Medical Services," 20 (n. 95).

108. Stone to M. D. Whitaker, 15 April 1943 (ANL Reading File); Parker to

Cantril et al., "Report on Contamination of Coveralls," 18 Nov. 1943 (ANL Chron File); Radiation Monitoring and Instrument Coordinating Committee to Stone, 15 May 1943 (n. 17); Warren, "Role of Radiology," 865–868 (n. 56); Merril Eisenbud to author, 6 March 1981.

109. Wirth, "Management and Treatment of Exposed Personnel," in Stone, *Industrial Medicine*, 266 (n. 9).

110. Wirth, "Medical Services," 33–34 (n. 95).

111. CH-2812, 24 (n. 23); Wirth, "Medical Services," 33–35 (n. 95); Eisenbud to author (n. 108).

112. Wirth, "Medical Services," 35 (n. 95).

113. Cantril, "Industrial Medical Program," 290 (n. 103).

114. Ibid., 289–307; Healy interview (n. 101); Cantril and Parker, "Status of Health and Protection at the Hanford Engineer Works," in Stone, ed., *Industrial Medicine*, 476–484 (n. 9).

115. Parker, "Protection Programs," 574 (n. 33).

116. Stone to Members of the Project Council et al., "Regarding Wearing Pocket Ionization Chambers and Other Devises [sic] for Determining Radiation Exposure," 4 Aug. 1943 (ANL Reading File); interview with Karl Z. Morgan, Atlanta, Ga., 25 July 1979. See also the reports of the Met Lab Chemistry Division Hazard Control Committee, such as CN-2408, "Product [Plutonium] Hazard Control Regulations and General Safety Precautions," 20 Nov. 1944; and CH-2493, "Rules for Fission Product Hazard Control and Recommendations for Operations with Fission Products," 19 Dec. 1944.

117. Wirth, "Medical Services," 32 (n. 95).

118. Warren, "Role of Radiology," 879 (n. 56).

119. Lombard Squires to J. N. Tilley and F. S. Chambers, "Packaging the Final Product for Storage and Shipment," 29 April 1944 (ANL Chron File).

120. Groves, *Now It Can Be Told*, 422 (n. 54).

121. Harley A. Wilhelm, "Development of Uranium Metal Production in America," *Journal of Chemical Education* 37 (Feb. 1960):59–67; Warren, "Role of Radiology," 868–870 (n. 56); Ferry interview (n. 7); Eisenbud to author (n. 108). The ANL Reading File contains numerous inspection reports.

122. Warren, "Role of Radiology," 869 (n. 56).

123. On the evidence of high levels of worker exposure to dust and fumes in uranium-processing facilities, albeit with no apparent serious ill effects, see Merril Eisenbud and J. A. Quigley, "Industrial Hygiene of Uranium Processing," *A.M.A. Archives of Industrial Health* 14 (July 1956):12–22; and Eisenbud, "Early Occupational Exposure Experience with Uranium Processing," in *Proceedings of the Conference on Occupational Health Experience with Uranium* (ERDA report 93 UC 41, 1976). John Walsh, "A Manhattan Project Postscript," *Science* 212 (19 June 1981):1369–1371, recalls his work and some of the hazards at a New Jersey plant making uranium early in the war. His recollection was prompted by the initiation of a massive cleanup of former project sites; see Jerry Knight, "U.S. Hunts A-Bomb Project Debris," *Los Angeles Times*, 20 Jan. 1981, sect. IV, p. 13.

124. Compton to Conant, 11 Jan. (n. 51); K. Z. Morgan to R. S. Stone, "The

Past and the Future Health-Physics Programs of Clinton Laboratories," 30 Aug. 1945, 3 (ANL Chron File); Compton, *Atomic Quest*, 179–181 (n. 53); Morgan, "Instrumentation," 76–79 (n. 12); Neal O. Hines, *Proving Ground: An Account of the Radiobiological Studies in the Pacific, 1946–1961* (Seattle: University of Washington Press, 1962), chap. 1; interviews with John R. Farmakes, Argonne, 20 June 1979; Jack Bailey, Oak Ridge, 19 July 1979; and Healy (n. 101).

125. Stone, "General Introduction," 16 (n. 20).

3. RADIATION SAFETY AT LOS ALAMOS

1. James B. Conant and Leslie R. Groves to J. Robert Oppenheimer, 25 Feb. 1943 (all documents cited, unless otherwise indicated, are from Mail and Records Center, Los Alamos Scientific Laboratory [LASL], Los Alamos, N.M.); interview with Louis H. Hempelmann, Jr., Los Alamos, N.M., 3–4 June 1980.

2. Hempelmann to Oppenheimer, 11 March 1943; Robert S. Stone to Emilio Segrè, 27 March 1943 (Metallurgical Laboratory Reading File, Argonne National Laboratory, Argonne, Ill.); Hempelmann interview (n. 1).

3. Hempelmann interview (n. 1); interview with James F. Nolan, Los Angeles, 17 April 1980; Nolan to Richard F. Newcomb, 12 Aug. 1957 (copy in Dosimetry Research Project, Reynolds Electrical & Engineering Co., Las Vegas, Nev.; hereafter cited as REECo files).

4. Marjorie Bell Chambers, "Technically Sweet Los Alamos: The Development of a Federally Sponsored Scientific Community," Ph.D. diss., History, University of New Mexico, 1974, part 1; Lawrence Badash, Joseph O. Hirschfelder, and Herbert P. Broida, eds., *Reminiscences of Los Alamos, 1943–1945* (Dordrecht, Holland: D. Reidel, 1980); James W. Kunetka, *City of Fire: Los Alamos and the Birth of the Atomic Age, 1943–1945* (Englewood Cliffs, N.J.: Prentice-Hall, 1978), chap. 2.

5. Bernice Brode, "Tales of Los Alamos," in Badash et al., eds., *Reminiscences of Los Alamos*, 139 (n. 4).

6. Nolan interview (n. 3); Nolan to Newcomb (n. 3); Kunetka, *City of Fire,* chap. 5 (n. 4); David Hawkins, "Manhattan District History, Project Y, the Los Alamos Project," vol. 1: "Inception until August 1945," LASL report LAMS-2532 (Vol. I), completed Aug. 1946, issued Dec. 1961, 4; Lenore Fine and Jesse A. Remington, *The Corps of Engineers: Construction in the United States,* United States Army in World War II: The Technical Services (Washington: Office of the Chief of Military History, 1972), 693–694. A slightly edited and repaginated version of Hawkins's study has been published as "Part I. Toward Trinity," in *Project Y: The Los Alamos Story,* History of Modern Physics, 1800–1950, vol. 2 (Los Angeles and San Francisco: Tomash, 1983).

7. Hempelmann, "History of the Health Group (A-6) (March 1943–November 1945)," April 1946, 1; Hawkins, "Project Y," 59 (n. 6); Hempelmann interview (n. 1); Nolan interview (n. 3); Nolan to Newcomb (n. 3).

8. Hempelmann, "Health Report," 9 Aug. 1943; Hawkins, "Project Y," 60 (n. 6); Hempelmann interview (n. 1); Laura Fermi, *Atoms in the Family: My Life with Enrico Fermi* (Chicago: University of Chicago Press, 1954), 227–28.

9. Hempelmann, "History," 3–4 (n. 7). Cf. Hempelmann, Joseph G. Hoffman, and Wright H. Langham, "Summary of Medical and Biological Research Activities of the Los Alamos Scientific Laboratory (1942 to 1947)," n.d., 1.

10. Stone to Segrè, 27 March (n. 2); Hempelmann to James Chadwick, "Recent Developments Dealing with Biological Effects of Radiation," 5 June 1944.

11. Hempelmann, "Health Report," 9 Aug. (n. 8).

12. Hempelmann, "History," 2–3 (n. 7); Hawkins, "Project Y," 60 (n. 6); Hempelmann interview (n. 1); Nolan interview (n. 3).

13. Hempelmann, "Health Report," 19 Jan. 1944.

14. Hempelmann to Oppenheimer, "Danger to Personnel from Inhaled or Ingested Plutonium," ca. 8 Feb. 1944.

15. Kunetka, City of Fire, 80–81 (n. 4).

16. Wright H. Langham and John W. Healy, "Maximum Permissible Body Burdens and Concentrations of Plutonium: Biological Basis and History of Development," in Harold C. Hodge, J. Newell Stannard, and J. B. Hursh, eds., Uranium, Plutonium, Transplutonic Elements, Handbook of Experimental Pharmacology, vol. 36 (New York: Springer-Verlag, 1973), 572; Herbert M. Parker to author, 10 March 1981 (REECo files). See chap. 2, above.

17. Cyril Stanley Smith to Kennedy, "Hazard and Insurance," 24 Feb. 1944.

18. David Dow to R. M. Underhill, 2 March 1944; Hawkins, "Project Y," 50–51 (n. 6).

19. Hempelmann to Oppenheimer, 8 Feb. 1944 (n. 14); E. R. Russell and James J. Nickson, "Distribution and Excretion of Plutonium," in Robert S. Stone, ed., Industrial Medicine on the Plutonium Project: Survey and Collected Papers, National Nuclear Energy Series, Div. IV, vol. 20 (New York: McGraw-Hill, 1951), 256.

20. Oppenheimer to Arthur H. Compton, 11 Feb. 1944.

21. Oppenheimer to Groves, 15 Feb. 1944; Langham and Healy, "Maximum Permissible Body Burdens," 572–574 (n. 16). See chap. 2, above.

22. Hawkins, "Project Y," 61, (n. 6); Russell and Nickson, "Distribution and Excretion of Plutonium," 256 (n. 19); Langham and Healy, "Maximum Permissible Body Burdens," 576 (n. 16); Raymond D. Finkle to author, 23 Feb. 1981 (REECo files).

23. Kenneth D. Nichols to Compton, 15 Sept. 1944, w/encl. Stafford L. Warren to Nichols, "Radiation Hazards," 1 Sept. 1944 (in Chronological File of Miscellaneous Documents [Historical Interest], Argonne National Laboratory; cited hereafter as ANL Chron File); Joseph G. Hamilton to Stone, 18 Oct. 1944, w/encl. "Estimated Tolerance Doses for Product [Plutonium]" (E. O. Lawrence Collection, Bancroft Library, University of California, Berkeley, carton 5, folder 9); Warren for files, "Medical Experimental Program on Radium and Product [Plutonium]," 11 May 1945 (Lawrence Coll., 28/41). See chap. 2, above.

24. Hempelmann, "History," 11 (n. 7); Hawkins, "Project Y," 185 (n. 6); A. Lacassagne, "Historical Outline of the Initial Studies on the Use of Polonium in Biology," British Journal of Radiology 44 (1971):546–458.

25. Kennedy to circulation list, "Control of Health Hazards," 24 Feb. 1944; Hempelmann to Warren, 31 March 1944; Hempelmann, "History," 5–6 (n. 7).

26. Oppenheimer to Samuel K. Allison, 22 Feb. 1944 (ANL Chron File).

27. Kennedy to circulation list, 24 Feb. (n. 25).

28. Allison to Oppenheimer, 1 March 1944; Hempelmann to Kennedy and Kenneth Bainbridge, "Effects of Toxic Agents," 12 April 1944; Hempelmann, "History," 8 (n. 7); "Explanatory Notes—Health-Safety Records," Los Alamos Health Division, ca. 1947.

29. William P. Jesse, "The Role of Instruments in the Atomic Bomb Project," *Chemical Engineering News* 24 (1946):2906–2909; Ronald L. Kathren, "Historical Development of Radiation Measurement and Protection," in Allen B. Brodsky, ed., *CRC Handbook of Radiation Measurement and Protection*, Section A, vol. 1: *Physical Science and Engineering Data* (West Palm Beach, Fla.: CRC Press, 1978), 48; Ralph E. Lapp, "Survey of Nucleonics Instrumentation Industry," *Nucleonics* 4 (May 1949):101–102. See chap 2, above.

30. Hempelmann to Jesse, 26 June 1944 (ANL Chron File); Paul C. Aebersold, Hempelmann, William H. Hinch, Joseph G. Hoffman, Wright H. Langham, and J. F. Tribby, "Chemistry and Metallurgy Health Handbook of Radioactive Materials," 17 Aug. 1945, 12–14.

31. Hempelmann, "History," 5, 9–10 (n. 7).

32. J. Garrott Allen to author, 14 Feb. 1981 (REECo files).

33. John E. Wirth, "Management and Treatment of Exposed Personnel," in Stone, ed., *Industrial Medicine*, 264–275 (n. 19).

34. Frederic de Hoffman, "Pure Science in the Service of Wartime Technology," *Bulletin of the Atomic Scientists* 31 (Jan. 1975):42.

35. Cyril Stanley Smith, "Plutonium Metallurgy at Los Alamos during 1943–45," in A. S. Coffinberry and W. N. Miner, eds., *The Metal Plutonium* (Chicago: University of Chicago Press, 1961), 26–35.

36. Hawkins, "Project Y," 263 (n. 6).

37. Hempelmann, "Health Report," 30 Aug. 1944; Hawkins, "Project Y," 256 (n. 6); Glenn T. Seaborg, "Plutonium Revisited," in Betsy J. Stover and Webster S. S. Lee, eds., *Radiobiology of Plutonium* (Salt Lake City: J. W. Press, Dept. of Anatomy, University of Utah, 1972), 17–19.

38. Hempelmann, "Health Report," 30 Aug. (n. 37).

39. Ibid.; Hempelmann, "History," 7 (n. 7); Aebersold et al., "Health Handbook" (n. 30).

40. Hempelmann, "Health Report," 30 Aug. (n. 37); Hempelmann, "History," 2 (n. 7); Warren for files, 11 May 1945 (n. 23); Langham and Healy, "Maximum Permissible Body Burdens," 575–576 (n. 16).

41. Hempelmann, "Health Report," Aug. (n. 37); Hempelmann, "History," 2 (n. 7).

42. Hempelmann to Oppenheimer, "Health Hazards Related to Plutonium," 16 Aug. 1944.

43. Oppenheimer to Hempelmann, "Your Memorandum of August 16, 1944," 16 Aug. 1944.

44. Minutes of the Meeting of the Administrative Board, 17 Aug. 1944.

45. Warren for files, 11 May (n. 23).

46. Hempelmann to Oppenheimer, "Medical Research Program," 29 Aug. 1944.

47. Donald F. Mastick, summary of proposed medical research program, 23

Aug. 1944; Hempelmann to Kennedy, "Advisory Committee for Medical Reseach Program," 29 Aug. 1944.

48. Hempelmann to Oppenheimer, "Meeting of Chemistry Division and Medical Group," 26 March 1945; Wright H. Langham to Hempelmann, "Report of Talk Given at the Chicago Meeting, May 16, 1945: The Monitoring of Site Y Personnel for 49 [Plutonium] Contamination," 18 June 1945; Wirth for medical files, "Present Status of Relationship of Analyses of Plutonium in Urine to the Plutonium Hazard," 28 May 1945 (ANL Chron File); Hempelmann, "History," 7–8 (n. 7); Hawkins, "Project Y," 184 (n. 6); Russell and Nickson, "Distribution and Excretion of Plutonium," 257–260 (n. 19); Hempelmann et al., "Summary of Medical and Biological Research," 5–8 (n. 9); Langham, Samuel H. Bassett, Payne S. Harris, and Robert E. Carter, "Distribution and Excretion of Plutonium Administered Intravenously to Man," LASL report no. LA-1151, 20 Sept. 1950 (published in Health Physics 38 [1980]:1031–1060); Langham and Healy, "Maximum Permissible Body Burdens," 576 (n. 16).

49. Hempelmann to Members of Research Group, "Future Experiments," 2 March 1945; Kennedy, A. C. Wahl, Nolan, Langham, and Hempelmann to Oppenheimer, "Medical Research of Manhattan District Concerned with Plutonium," 15 March 1945; Hempelmann to Oppenheimer, 29 Aug. (n. 46); Oppenheimer to Warren, 29 March 1945; Warren to Oppenheimer, 15 May 1945; Wirth, "Management and Treatment of Exposed Personnel," 274 (n. 33); Langham and Healy, "Maximum Permissible Body Burdens," 576 (n. 16); George L. Voelz, Hempelmann, J. N. P. Lawrence, and William D. Moss, "A 32-Year Follow-up of Manhattan Project Plutonium Workers," Health Physics 37 (1979): 445–485.

50. Hempelmann, "History," 15–16 (n. 7).

51. Richard G. Hewlett and Oscar E. Anderson, Jr., A History of the United States Atomic Energy Commission, vol. 1: The New World, 1939/1946 (University Park: Pennsylvania State University Press, 1962), 235, 245–249, 251–252, 311.

52. Hempelmann to Robert H. Dunlap, "Health Safety Program," 10 Oct. 1944; Dunlap to Kennedy, "Relationship CM-1 and A-6," 27 Jan. 1945; Hempelmann to Kennedy, "Memorandum of R. H. Dunlap, 27 January 1945," 28 Jan. 1945; Hempelmann, "History," 10 (n. 7).

53. William H. Hinch to Hempelmann, "Split of Responsibilities of H. I. [Health Instruments] and Medical Groups in the CM Division," 23 April 1945; Hempelmann to Kennedy, "Exposure of Personnel in Recovery Group," 23 June 1945; R. W. Dodson to G. Friedlander, L. Helmholz, and I. Johns, "Radiation Health Hazards," 11 July 1945; Hempelmann, "History," 10–11 (n. 7); Hawkins, "Project Y," 184–185, 263–265 (n. 6).

54. Hempelmann, "History," 1 (n. 7).

55. Ibid., 4–5.

56. Hawkins, "Project Y," 116–117, 185–186, 218 (n. 6); Hempelmann, "History," 12 (n. 7).

57. L. D. P. King, "Omega Health Hazards, Group F-2," n.d. Cf. Hempelmann, "History," 12 (n. 7).

58. Hempelmann "Health Report," 30 Aug. (n. 37); Hempelmann for files, "The Second Ra-La Experiment (4 October 1944)," 12 Oct. 1944; Hawkins, "Project Y," 142, 234–235, 251 (n. 6).

59. R. W. Dodson to Hempelmann, "Gamma Ray Dosages," 19 Feb. 1945; Hawkins, "Project Y," 259–260 (n. 6).

60. Hempelmann, "History," 12 (n. 7); Hawkins, "Project Y," 186 (n. 6).

61. Margaret Gowing, *Britain and Atomic Energy, 1939–1945* (London: Macmillan; New York: St. Martin's Press, 1964); Ronald W. Clark, *The Birth of the Bomb* (New York: Horizon Press, 1961).

62. Otto R. Frisch, "Somebody Turned the Sun on with a Switch," *Bulletin of the Atomic Scientists* 30 (April 1974):18; Frisch, "The Los Alamos Experience," *New Scientist* 83 (19 July 1979):187; Frisch, *What Little I Remember* (Cambridge: Cambridge University Press, 1979), 159–160; Hewlett and Anderson, *New World*, 317 (n. 51); Hawkins, "Project Y," 229–230 (n. 6).

63. Frisch, "Los Alamos Experience," 187 (n. 62).

64. As quoted by Frisch, *What Little I Remember*, 159 (n. 62). Cf. Hawkins, "Project Y," 198 (n. 6).

65. Frisch, "Los Alamos Experience," 187 (n. 62).

66. Hoffman, "Pure Science," 44 (n. 34).

67. Frisch, "Los Alamos Experience," 187 (n. 62); Hempelmann, "History," 12 (n. 7).

68. Hawkins, "Project Y," 230 (n. 6).

69. Col. A. W. Betts to Groves, "Radiation Accident at Site Y," 27 May 1946; Hempelmann, "History," 12–13 (n. 7); Frisch, *What Little I Remember*, 160–161 (n. 62); Edith C. Truslow and Ralph Carlisle Smith, "Manhattan District History, Project Y, the Los Alamos Project," vol. 2: "August 1945 through December 1946," LASL report no. LAMS-2532 (Vol. II), completed 1947, issued 1961, 17, 50–51, 76–77 (slightly edited and repaginated version published as Truslow and Smith, "Part II. Beyond Trinity," in *Project Y* [n. 6]); Hempelmann, Hermann Lisco, and Joseph G. Hoffman, "The Acute Radiation Syndrome: A Study of Nine Cases and a Review of the Problem," *Annals of Internal Medicine* 36 (1952):279–500; Leona Marshall Libby, *The Uranium People* (New York: Crane Russak and Charles Scribner's Sons, 1979), 202–204.

70. Kenneth T. Bainbridge, *Trinity*, LASL report no. LA-6300-H (Los Alamos, May 1976), 1–2 (this declassified, published version contains almost all the original text of the secret report LA-1012 which Bainbridge, the test director, wrote in 1946); Bainbridge, "Prelude to Trinity," *Bulletin of the Atomic Scientists* 31 (April 1975):42–44; George B. Kistiakowsky, "Reminiscences of Wartime Los Alamos," in Badash et al., eds., *Reminiscences of Los Alamos*, 53 (n. 4).

71. Kistiakowsky to Oppenheimer, "Activities at Trinity," 13 Oct. 1944; Oppenheimer to Bainbridge, "Test Measurements Program for Trinity," 15 March 1945; Bainbridge, *Trinity*, 4–5 (n. 70); Bainbridge, "Prelude," 44 (n. 70); Kistiakowsky, "Reminiscences," 55–56 (n. 70).

72. Kistiakowsky to Oppenheimer, 13 Oct. (n. 71); P. B. Moon to J. H. Manley, "Plans for Measurements of Nuclear Radiations at Trinity Test," 14 Feb. 1945; Lansing Lamont, *Day of Trinity* (New York: Atheneum, 1965), 158, 189;

Herbert L. Anderson, "Fermi, Szilard and Trinity," *Bulletin of the Atomic Scientists* 30 (Oct. 1974):46.

73. Bainbridge, "Prelude," 44 (n. 70).

74. Bainbridge, *Trinity*, 3 (n. 70); Hawkins, "Project Y," 267 (n. 6); Bainbridge, "Prelude," 44–46 (n. 70); Lamont, *Day of Trinity*, 64–65 (n. 72); Kunetka, *City of Fire*, 146–147 (n. 4).

75. Oppenheimer to Groves, 20 Oct. 1962, as quoted in Alice Kimball Smith and Charles Weiner, eds., *Robert Oppenheimer: Letters and Recollections* (Cambridge, Mass., and London: Harvard University Press, 1980), 290; Chambers, "Technically Sweet Los Alamos," 142 (n. 4). See also Ferenc Morton Szasz, *The Day the Sun Rose Twice: The Story of the Trinity Site Nuclear Explosion, July 16, 1945* (Albuquerque: University of New Mexico Press, 1984), 40–41.

76. Kistiakowsky to Oppenheimer, 13 Oct. (n. 71); Bainbridge, *Trinity*, 4–8 (n. 70); Hawkins, "Project Y," 268–270 (n. 6); Hewlett and Anderson, *New World*, 311–320 (n. 51); Bainbridge, "A Foul and Awesome Display," *Bulletin of the Atomic Scientists* 31 (May 1975):40.

77. Hempelmann, "History," 4–5 (n. 7).

78. Leslie R. Groves, "Some Recollections of July 16, 1945," *Bulletin of the Atomic Scientists* 26 (June 1970):24.

79. "Report by M.A.U.D. Committee on the Use of Uranium for a Bomb," July 1941, repr. in Gowing, *Britain and Atomic Energy*, 407 (n. 61).

80. Committee on Biological Effects of the Gadget to Oppenheimer, "Priority on Further Investigation of Problem," 22 May 1944.

81. Hempelmann, "Hazards of Trinity Experiment, Section I," 12 April 1945.

82. Ibid.; Hempelmann, "History of the Preparation of the Medical Group for Trinity Test II," n.d., in Hempelmann, "Preparation and Operational Plan of Medical Group (TR-7) for Nuclear Explosion 16 July 1945," LASL report LA-631, 13 June 1947, 4.

83. Hempelmann, "Hazards" (n. 81).

84. Ibid.

85. Ibid.; Samuel Glasstone, ed., *The Effects of Atomic Weapons* (Washington: Government Printing Office, 1950), 31–35, 270–274.

86. Hempelmann, "Hazards" (n. 81).

87. Richard J. Watts, "Hazards of Trinity Experiment, Section II," 12 April 1945.

88. Watts, "Hazards" (n. 87); Watts, "Health Instrumentation and Crater Activity Decay for the July 16th Nuclear Explosion," App. I in Joseph G. Hoffman, "Nuclear Explosion 16 July 1945: Health Physics Report on Radioactive Contamination throughout New Mexico Following the Nuclear Explosion, Part A: Physics," LASL report LA 626, 20 Feb. 1947, 67–68, 72–73.

89. Watts, "Hazards" (n. 87).

90. Watts, "Hazards" (n. 87); Watts, "Health Instrumentation," 69 (n. 88); Aebersold et al., "Health Handbook," 12–13 (n. 30); Glasstone, ed., *Effects of Atomic Weapons*, 294–295, 300–301 (n. 85).

91. Watts, "Hazards" (n. 87); Glasstone, ed., *Effects of Atomic Weapon*, 303 (n. 85); Jesse, "Role of Instrumentation," 2908 (n. 29).

92. Watts, "Health Instrumentation," 69–70, 72–73 (n. 88); Aebersold et al., "Health Handbook," 14 (n. 30).

93. Watts, "Hazards" (n. 87); Watts, "Health Instrumentation," 70–71 (n. 88); Kathren, "Historical Development," 48 (n. 29); Richard D. Terry, "Historical Development of Commercial Health Physics Instrumentation," in Kathren and Paul L. Ziemer, eds., *Health Physics, A Backward Glance. Thirteen Original Papers on the History of Radiation Protection* (New York: Pergamon Press, 1980), 160–161.

94. Watts, "Health Instrumentation," 71 (n. 88); Hoffman, "Health Physics Report," 7 (n. 88); Glasstone, ed., *Effects of Atomic Weapons*, 301–304, 307 (n. 85).

95. Hoffman, "Health Physics Report," 7 (n. 88).

96. Watts, "Hazards" (n. 87); Glasstone, ed., *Effects of Atomic Weapons*, 304–305 (n. 85).

97. Watts, "Health Instrumentation," 71 (n. 88).

98. Watts, "Hazards" (n. 87).

99. Hempelmann, "Hazards" (n. 81); Watts, "Hazards" (n. 87); Hempelmann, "Preparation," 4 (n. 82).

100. H. L. Anderson to Bainbridge, "The 100 Ton Shot Preparations—Addition no. 7 to Project TR Circular," 19 April 1945; Hempelmann for files, "Hazards of 100 Ton Shot at Trinity," 18 May 1946, in Hempelmann, "Preparation," 88–89 (n. 82); Hawkins, "Project Y," 270–271 (n. 6).

101. Anderson to Bainbridge, 19 April (n. 100); Hempelmann for files, 18 May (n. 100).

102. Dan Mayers, "Health Hazards, Group F-4," n.d., 1.

103. R. C. Tolman to Groves, "First Trinity Test," 13 May 1945, in Bainbridge, *Trinity*, 9–13 (n. 70); Hempelmann for files, 18 May (n. 100).

104. Hempelmann for files, 18 May (n. 100).

105. Ibid.

106. Warren to Groves, "Visit to Sites M and Y on 6–10 May, incl., 1945 during Test I," 16 May 1945, par. 5.

107. Anderson to Bainbridge, 19 April (n. 100); Tolman to Groves, 13 May (n. 103); Hempelmann for files, 18 May (n. 100).

108. Anderson to Bainbridge, 19 April (n. 100); Tolman to Groves, 13 May (n. 103).

109. Bainbridge to Hubbard, 19 April 1945; Anderson to Bainbridge, "Meteorological Aspects of Sampling," 24 April 1945; Hubbard, "Meteorological Report Concerning the May 7th Operation at Trinity," 7 May 1945.

110. Hempelmann for files, 18 May (n. 100).

111. Tolman to Groves, 13 May (n. 103); Hawkins, "Project Y," 273 (n. 6).

112. Tolman to Groves, 13 May (n. 103); Hempelmann, "Preparation," 4 (n. 82).

113. Hawkins, "Project Y," 271 (n. 6).

114. Samuel K. Allison to Oppenheimer, "Discussion of Meteorological Effects Following Shots," 10 May 1945; Bainbridge to All Concerned, "Notes on Meeting of May 21, 1945," 31 May 1945; Groves, "Recollections," 24 (n. 78).

4. TRINITY

1. Leslie R. Groves, *Now It Can Be Told: The Story of the Manhattan Project* (New York and Evanston: Harper & Row, 1962), chap. 10; Groves, "Some Recollections of July 16, 1945," *Bulletin of the Atomic Scientists* 26 (June 1970):24; Groves, "The A-Bomb Program," in Fremont E. Kast and James E. Rosenzweig, eds., *Science, Technology, and Management* (New York: McGraw-Hill, 1963), 31–40; Richard G. Hewlett and Oscar E. Anderson, Jr., *A History of the United States Atomic Energy Commission*, vol. 1: *The New World, 1939/1946* (University Park: Pennsylvania State University Press, 1962), 227–229; Martin J. Sherwin, *A World Destroyed: The Atomic Bomb and the Grand Alliance* (New York: Alfred A. Knopf, 1975), 58–63.

2. James H. Nolan to Richard F. Newcomb, 12 Aug. 1957 (REECo files).

3. Interview with Hymer L. Friedell, Cleveland, 12 June 1979.

4. Groves, "Recollections," 24 (n. 1).

5. Ibid.

6. Groves to J. Robert Oppenheimer, 27 April 1945 (all documents, unless otherwise indicated, are in the Mail and Records Center, Los Alamos Scientific Laboratory [LASL], Los Alamos, N.M.).

7. Kenneth T. Bainbridge to Capt. T. O. Jones, "Legal Aspects of TR Tests," 2 May 1945.

8. Groves to Oppenheimer, 27 April (n. 6).

9. Oppenheimer to Groves, 27 June 1945.

10. Groves, "Recollections," 24 (n. 1); David Hawkins, "Manhattan District History, Project Y, the Los Alamos Project," vol. 1: "Inception until August 1945," LASL report LAMS-2532 (Vol. I), completed Aug. 1946, issued Dec. 1961, 272; Lansing Lamont, *Day of Trinity* (New York: Atheneum, 1965), 158–159.

11. Hawkins to Lamont, 28 Sept. 1964 (Lansing Lamont Papers, Harry S. Truman Library, Independence, Mo.; cited hereafter as Lamont Papers); Lamont, *Day of Trinity*, 129–130 (n. 10).

12. Kenneth T. Bainbridge, "A Foul and Awesome Display," *Bulletin of the Atomic Scientists* 31 (May 1975):44.

13. Francis William Aston, *Isotopes* (London: Edward Arnold, 1922), as quoted in Bainbridge, "A Foul and Awesome Display," 44 (n. 12). Cf. Brian Easlea, *Fathering the Unthinkable: Masculinity, Scientists and the Nuclear Arms Race* (London: Pluto Press, 1983), 49–58, especially at 52–53.

14. Val Fitch, "The View from the Bottom," *Bulletin of the Atomic Scientists* 31 (Feb. 1975):44.

15. Jeremy Bernstein, *Hans Bethe: Prophet of Energy* (New York: Basic Books, 1980), 73; Stanley A. Blumberg and Gwinn Owens, *Energy and Conflict: The Life and Times of Edward Teller* (New York: G. P. Putnam's Sons, 1976), 116–119. Cf. Ferenc Morton Szasz, *The Day the Sun Rose Twice: The Story of the Trinity Site Nuclear Explosion, July 16, 1945* (Albuquerque: University of New Mexico Press, 1984), 56–60.

16. Bainbridge, "A Foul and Awesome Display," 44 (n. 12).

17. Groves, *Now It Can Be Told*, 298 (n. 1).

18. William L. Laurence, *Men and Atoms: The Discovery, the Uses and the Future of Atomic Energy* (New York: Simon & Schuster, 1959), 10–11.

19. Groves, "Recollections," 24–25 (n. 1).

20. Bainbridge to J. M. Hubbard, 19 April 1945; Joseph O. Hirschfelder to Oppenheimer, "Strategic Possibilities Arising if a Thunderstorm Is Induced by Gadget Explosion," 25 April 1945; Samuel K. Allison to Oppenheimer, "Discussion of Meteorological Effects Following Shots," 10 May 1945; Stafford L. Warren to Groves, "Analysis of the Problems Presented by Test II at Muriel [Alamogordo]," 16 May 1945; P. E. Church to Oppenheimer, 18 May 1945; Bainbridge to All Concerned, "Notes on Meeting of May 28," 31 May 1945; Louis H. Hempelmann, Jr., "Preparation and Operational Plan of Medical Group (TR-7) for Nuclear Explosion 16 July 1945," LASL report no. LA-631 (Los Alamos, 13 June 1947), 4.

21. Bainbridge to Jones, 2 May (n. 7); Church to Oppenheimer, 18 May (n. 20); Hubbard to Bainbridge, "The Operational Plan of TR4 through June and July," 26 May 1945; Paul C. Aebersold, "July 16th Nuclear Explosion: Safety and Monitoring of Personnel," LASL report no. LAMS-616 (Los Alamos, 25 Jan. 1946), 17; Hawkins, "Project Y," 273 (n. 10); Groves, "Recollections," 24 (n. 1); Lamont, *Day of Trinity*, 159–160 (n. 10); Stafford L. Warren, "The Role of Radiology in the Development of the Atomic Bomb," in Kenneth D. A. Allen, ed., *Radiology in World War II*, Medical Department in World War II: Clinical Studies (Washington: Office of the Surgeon General, U.S. Army, 1966), 883.

22. Oppenheimer to Groves, 27 June (n. 9).

23. Bainbridge to Washington Liaison Office, 2 July 1945. Cf. Warren to Groves, "Safeguards for Test II at Muriel," 27 June 1945.

24. Nolan to John Williams, "Reference Your TR Memorandum 17 May 1945," 18 May 1945; Bainbridge to All Concerned, 31 May (n. 20); Bainbridge to All Concerned, "Project TR Organization," TR Circular no. 10, 1 June 1945; Aebersold to Hempelmann, "TR Site Monitoring Plans as of July 11, 1945 (Supplement to Medical Hazards of TR #2 by Capt. Nolan)," 11 July 1945; Hempelmann, "Preparation," 4 (n. 20); interview with James F. Nolan, Los Angeles, 17 April 1980.

25. Nolan, "Medical Hazards of TR #2," 20 June 1945. A version of Nolan's memo appears under Hempelmann's name in Bainbridge, *Trinity*, LASL report no. LA-6300-H (Los Alamos, May 1976), 31–36.

26. Nolan, "Medical Hazards" (n. 25); Aebersold to Hempelmann, 11 July (n. 24).

27. Allison to Oppenheimer, 10 May (n. 20).

28. Groves, "Recollections," 24 (n. 1).

29. Bainbridge to All Concerned, 31 May (n. 20).

30. Aebersold to Hempelmann, 11 July (n. 24); "To People Entering the Area after the Shot," n.d., a set of rules to be read and signed before entry; Aebersold, "Safety and Monitoring," 25–27 (n. 21); Bainbridge, *Trinity*, 16 (n. 25).

31. Herbert L. Anderson to Bainbridge, "Operational Plans for Trinity," 14

June 1945; Anderson, "Fermi, Szilard and Trinity," *Bulletin of the Atomic Scientists* 30 (Oct. 1974):46.

32. Bainbridge to All Concerned, 31 May (n. 20); Aebersold to Hempelmann, 11 July (n. 24); "Abstract of Conferences on Medical Hazards Held before Trinity," 10 Aug. 1945, entry for 15 July; Aebersold, "Safety and Monitoring," 29–30 (n. 21).

33. "To People Entering" (n. 30).

34. Groves to Oppenheimer, 27 April (n. 6).

35. Nolan, "Medical Hazards" (n. 25); Hempelmann, "Preparation," 4 (n. 20); Hawkins, "Project Y," 272 (n. 10).

36. Bainbridge to Jones, 2 May (n. 7).

37. Nolan interview (n. 24).

38. Joseph O. Hirschfelder and John L. Magee to Bainbridge, "Danger from Active Material Falling from Cloud," 16 June 1945; Hirschfelder, "The Scientific and Technological Miracle at Los Alamos," in Lawrence Badash, Hirschfelder, and Herbert P. Broida, eds., *Reminiscences of Los Alamos, 1943–1945* (Dordrecht, Holland: D. Riedel, 1980), 72–75. Cf. Hubbard to Bainbridge, 26 May 1945; N. F. Ramsey to Oppenheimer, "Observational Equipment," 9 June 1945.

39. Messrs. Penney, Reines, and Taylor to Messrs. Bainbridge, Oppenheimer, and Williams, "Contamination by Dust Particles: Preventive Measures in the Vicinity of Pt. O," 20 June 1945.

40. Hempelmann and Nolan to Bainbridge, "Danger to Personnel in Nearby Towns Exposed to Active Material Falling from Cloud," 22 June 1945.

41. Hirschfelder and Magee, "Improbability of Danger from Active Material Falling from Cloud," 6 July 1945; Aebersold notes on "Conference about Contamination of Countryside near Trinity with Radioactive Materials," 10 July 1945; "Abstract of Conferences," entry for 10 July (n. 32).

42. Hubbard to Oppenheimer, "Long Range Weather Schedule," 21 June 1945; interview with General Leslie Groves by Lansing Lamont, 22 Aug. 1963 (Lamont Papers); Hawkins, "Project Y," 273 (n. 10); Bainbridge, *Trinity,* 28 (n. 25); Hewlett and Anderson, *New World,* chap. 11 (n. 1); Groves, "Recollections," 26–27 (n. 1); Bainbridge, "A Foul and Awesome Display," 43 (n. 12); Szasz, *Day the Sun Rose Twice,* 69–73 (n. 15).

43. Hempelmann, "Preparation," 5 (n. 20).

44. Brig. Gen. Thomas F. Farrell for file, "Assistance of 8th Service Command in Evacuation Plans," 26 June 1945; 1st Lt. Daniel H. Dailey to Office in Charge, "Trinity," 30 June 1945; Frank Oppenheimer and Bainbridge to Dailey, "Your Memo on Trinity of June 30," 30 June 1945; Bainbridge to Capt. T. O. Jones, "Memorandum of Discussion July 2," 3 July 1945; "Conference with Hoffman, Col. Warren, Major Palmer, and Hempelmann to Decide on Evacuation of Fifty Miles of Clouds Path (on the Basis of a N.E. Blow)," 10 July 1945, in Hempelmann, "Preparation," 43 (n. 20); Palmer, "Evacuation Detachment at Trinity," ibid., 48–50; "Abstract of Conferences," entry for 12 July (n. 32); Joseph G. Hoffman, "Nuclear Explosion 16 July 1945: Health Physics Report on Radioactive Contamination throughout New Mexico Following the Nuclear Ex-

plosion, Part A: Physics," LASL report no. LA-626 (Los Alamos, 20 Feb. 1947), 7–8; Hawkins, "Project Y," 272 (n. 10).

45. Warren to Groves, 27 June (n. 23); Warren to Groves, "Report on Test II at Trinity, 16 July 1945," 21 July 1945.

46. Friedell interview (n. 3).

47. Hoffman to Lt. D. Daley, "Town Monitoring," 5 July 1945; Hoffman, "Changes and Supplement to Town Monitoring," 7 July 1945; "Conference on Evacuation" (n. 44); Hoffman, "Town Monitoring Crew, Final Instructions," 10 July 1945; "Abstract of Conferences," entry for 14 July (n. 32); Hoffman, "Final Plans for Monitoring and Evacuation N.E. and N.W. Regions," 14 July 1945; Hempelmann to Bainbridge, "The Influence of Meteorologic Conditions on the Monitoring and Evacuation Plans of the Medical Group," 14 July 1945; Hempelmann, "Final Organization for Trinity Test #2," 15 July 1945; Hoffman, "Health Physics Report," 6–8 (n. 44); Hempelmann, "Preparation," 5 (n. 20).

48. Hoffman, "Final Plans" (n. 47).

49. Hoffman, "Health Physics Report," 7–8 (n. 44).

50. Hempelmann to Bainbridge, 14 July (n. 47).

51. Hempelmann to Hoffman, "Procedure to be Used by Town Monitors," 10 July 1945.

52. Hempelmann to Hoffman, 10 July (n. 51); Hoffman, "Instructions for Monitors," 10 July 1945; Hoffman, "Final Plans" (n. 47).

53. Hempelmann and Nolan to Bainbridge, 22 June (n. 40).

54. Aebersold notes, 14 July (n. 41). Cf. Warren to Groves, 27 June (n. 23); "Abstract of Conferences," entry for 10 July (n. 32).

55. Hoffman, "Health Physics Report," 18 (n. 44).

56. "Abstract of Conferences," entry for 14 July (n. 32).

57. Nolan to Newcomb (n. 2).

58. Nolan interview (n. 24); Hempelmann, "Preparation," 5 (n. 20).

59. "Abstract of Conferences," entry for 12 July (n. 32); Aebersold, "Safety and Monitoring," 8–9 (n. 21); Boyce McDaniel, "A Physicist at Los Alamos," *Bulletin of the Atomic Scientists* 30 (Dec. 1974):42–43; Hewlett and Anderson, *New World,* 378 (n. 1).

60. McDaniel, "Physicist at Los Alamos," 43 (n. 59).

61. W. F. Schaffer, "Work Preceding and Including Assembly at Trinity," in Bainbridge, *Trinity,* 40 (n. 25).

62. "Abstract of Conferences," entries for 12, 13 July (n. 32); oral comments on draft by Hempelmann, 6 Dec. 1984.

63. George B. Kistiakowsky, "Reminiscences of Wartime Los Alamos," in Badash et al., eds., *Reminiscences of Los Alamos,* 58 (n. 38).

64. Schaffer, "Work" (n. 61); Bainbridge, *Trinity,* 28–29 (n. 25).

65. "Abstract of Conferences," entry for 13 July (n. 32).

66. Hawkins, "Project Y," 274 (n. 10); McDaniel, "Physicist at Los Alamos," 43 (n. 59); Hewlett and Anderson, *New World,* 378 (n. 1).

67. Aebersold to Hempelmann, 11 July (n. 24).

68. Aebersold, "Safety and Monitoring," 9–10 (n. 21).

69. McDaniel, "Physicist at Los Alamos," 43 (n. 59). Cf. Hawkins, "Project Y," 274 (n. 10).

70. Hawkins, "Project Y," 274 (n. 10); Bainbridge, *Trinity*, 28 (n. 25); Hewlett and Anderson, *New World*, 378 (n. 1).

71. Aebersold, "Safety and Monitoring," 10 (n. 21).

72. Norris E. Bradbury, "TR Hot Run," 7 July 1945, in Bainbridge, *Trinity*, 43 (n. 25).

73. "Abstract of Conferences," passim (n. 32); Warren to Groves, 16 May (n. 20); Warren to Groves, 27 June (n. 23); Warren, "Role of Radiology," 881–884 (n. 21); interviews with Stafford L. Warren, Los Angeles, 30 Oct. 1979; and Louis H. Hempelmann, Jr., Los Alamos, 3–4 June 1980.

74. Howard C. Bush, "Memorandum to Camp Personnel," 14 July 1945; Bush and Warren, "Directions for Personnel at Base Camp at Time of Shot," 15 July 1945; "Directions for Personnel at Campana Hills Camp at Time of Shot (Coordinating Council Camp)," 15 July 1945; "Abstract of Conferences," entry for 15 July (n. 32); Aebersold, "Safety and Monitoring," 13, 92–94 (n. 21).

75. "Abstract of Conferences," entry for 14 July (n. 32); Hoffman, "Final Plans" (n. 47); Hempelmann to Bainbridge, 14 July (n. 47).

76. Hoffman, "Health Physics Report," 6–8 (n. 44); Hirschfelder, "Scientific and Technological Miracle," 75 (n. 38).

77. Hubbard, "Weather Forecast," 15 July 1945; "Abstract of Conferences," entry for 15 July, 8:00 P.M. (n. 32); Aebersold, "Safety and Monitoring," 10 (n. 21); McDaniel, "Physicist at Los Alamos," 43 (n. 59); Bainbridge, "A Foul and Awesome Display," 45 (n. 12).

78. Groves, "Recollections," 26 (n. 1).

79. Ibid.

80. Bainbridge, "A Foul and Awesome Display," 44 (n. 12).

81. "Abstract of Conferences," entry for 16 July, 2 A.M. (n. 32).

82. McDaniel, "Physicist at Los Alamos," 43 (n. 59).

83. Bainbridge, *Trinity*, 30 (n. 25); Hawkins, "Project Y," 275 (n. 10); Bainbridge, "A Foul and Awesome Display," 45 (n. 12); Groves, "Recollections," 26 (n. 1); Szasz, *Day the Sun Rose Twice*, 73–78 (n. 15).

84. Bainbridge, "A Foul and Awesome Display," 45–46 (n. 12).

85. "Abstract of Conferences," entry for 16 July, 5:00 A.M. (n. 32); Aebersold, "Safety and Monitoring," 10 (n. 21); Hawkins, "Project Y," 275 (n. 10); Bainbridge, *Trinity*, 30 (n. 25).

86. Bainbridge, *Trinity*, 31 (n. 25).

87. Bainbridge, "A Foul and Awesome Display," 46 (n. 12). For other examples, see Fitch, "View from the Bottom," 43 (n. 14); McDaniel, "Physicist at Los Alamos," 43 (n. 59); Hewlett and Anderson, *New World*, 379 (n. 1); Hawkins, "Project Y," 275–276 (n. 10); Szasz, *Day the Sun Rose Twice*, 83–91 (n. 15); Lamont, *Day of Trinity*, 180–187 (n. 10); James W. Kunetka, *City of Fire: Los Alamos and the Birth of the Atomic Age, 1943–1945* (Englewood Cliffs, N.J.: Prentice-Hall, 1978), 168–170; William L. Laurence, *Dawn over Zero: The Story of the Atomic Bomb* (2d ed.; New York: Alfred A. Knopf, 1946; repr. Westport, Conn.:

Greenwood Press, 1972), 10–12; O. R. Frisch, "Eye Witness Report of Nuclear Explosion, 16.7.45," as reproduced in Margaret Gowing, *Britain and Atomic Energy, 1939–1945* (London: Macmillan; New York: St. Martin's Press, 1964), 441–442.

88. Groves to Secretary of War, "The Test," 18 July 1945.

89. Bainbridge, "A Foul and Awesome Display," 46 (n. 12).

90. As quoted in Robert Jungk, *Brighter than a Thousand Suns: A Personal History of the Atomic Scientists,* trans. James Cleugh (New York: Harcourt, Brace, 1958), 201. See *Bhagavadgita* xi.32; Peter Goodchild, *J. Robert Oppenheimer: Shatterer of Worlds* (Boston: Houghton Mifflin, 1981), 162.

91. As quoted in Hewlett and Anderson, *New World,* 379 (n. 1).

92. Groves, "Recollections," 27 (n. 1); Aebersold to Hemplemann, 11 July (n. 24); Aebersold, "Safety and Monitoring," 16–17, 20 (n. 21).

93. "Events in Camp Immediately Following Shot—July 16, 1945 (Summarized from Col. Warren's and Hempelmann's Personal Notes)," in Hempelmann, "Preparation," 11 (n. 20); Aebersold, "Safety and Monitoring," 20, 28, 36, 37 (n. 21).

94. Victor Weisskopf, Hoffman, Aebersold, and Hempelmann to George Kistiakowsky, "Measurement of Blast, Radiation, Heat and Light and Radioactivity at Trinity," 5 Sept. 1945; Aebersold, "Safety and Monitoring," 20–25 (n. 21).

95. Aebersold to Hempelmann, "Protection and Monitoring on TR Site for Shot No. 2: Preliminary Report," 20 July 1945.

96. Bainbridge, "A Foul and Awesome Display," 46 (n. 12).

97. Weisskopf et al. to Kistiakowsky, 5 Sept. (n. 94); "Events in Camp," 11 (n. 93); Hempelmann to Groves, 25 Aug. 1945; Aebersold, "Safety and Monitoring," 38, 44 (n. 21).

98. Weisskopf et al. to Kistiakowsky, 5 Sept. (n. 94).

99. Interview with Herbert L. Anderson by Lansing Lamont, n.d. (Lamont Papers).

100. Aebersold, "Safety and Monitoring," 44–45 (n. 21). For a brief sketch of exposures recorded by Trinity participants, see Carl Maag and Steve Rohrer, *Project Trinity, 1945–1946,* report no. DNA 6028F (Washington: Defense Nuclear Agency, 15 Dec. 1982), chap. 4, "Dosimetry Analysis of Participants in Project Trinity,"

101. Aebersold, "Safety and Monitoring," 43–44, 45–46 (n. 21).

102. Ibid., 46–47.

103. "Events in Camp," 11 (n. 93); John Blair, Max Kupferberg, and Alex Nedzel to Hoffman, "Cloud Observations and Radiation Measurements," 20 July 1945; Warren to Groves, 21 July (n. 45); Bainbridge, "A Foul and Awesome Display," 46 (n. 12).

104. Hirschfelder and Magee to Bainbridge, "Fate of the Active Material after the Trinity Shot," 14 Aug. 1945.

105. Hoffman, "Health Physics Report," 8 (n. 44); Aebersold to Hempelmann, 20 July (n. 95); interview with Wright H. Langham by Lansing Lamont, n.d. (Lamont Papers).

106. Hirschfelder, "Scientific and Technological Miracle," 77 (n. 38).

107. Groves to Secretary of War, 18 July (n. 88); Aebersold, "Safety and Monitoring," 24 (n. 21); Hoffman, "Health Physics Report," 9 (n. 44).

108. Hirschfelder, "Scientific and Technological Miracle," 77 (n. 38).

109. Field notes of Trinity monitors, "L 8 Search Light Crew," 16 July 1945.

110. "Events in Camp," 11 (n. 93); Warren to Groves, 21 July (n. 45); Hoffman, "Health Physics Report," 8–10, 12–13, 28–29 (n. 44); Hirschfelder, "Scientific and Technological Miracle," 77–78 (n. 38); Weisskopf et al. to Kistiakowsky, 5 Sept. (n. 94); Hempelmann comments (n. 62).

111. "Events in Camp," 12–13 (n. 93); Palmer, "Evacuation" (n. 44); Hoffman, "Health Physics Report," 10–11 (n. 44).

112. Field notes of Alvin and Elizabeth Graves at Carrizozo, 16–17 July 1945; "Report of Mr. & Mrs. Allen [sic] Graves at Carrizozo, 16 July 1945," n.d.; Hoffman, "Health Physics Report," 11 (n. 44); Watts, "Health Instrumentation and Crater Activity Decay for the July 16th Nuclear Explosion," App. I, ibid, 74; Warren, "Role of Radiology," 885 (n. 21); Groves, "Recollections," 27 (n. 1); Robert Cahn, "Behind the First A-Bomb," *Saturday Evening Post*, 16 July 1960, 74.

113. "Events in Camp," 13–14 (n. 93).

114. Hempelmann, "Estimated Dose of Radiation Received by the Family in the Hot Canyon," n.d., in Hempelmann, comp., "Nuclear Explosion 16 July 1945: Health Physics Report on Radioactive Contamination throughout New Mexico, Part B: Biological Effects," Los Alamos, ca. April 1946; Hempelmann for Trinity files, "Follow-Up of Outlying Area Contaminated by Trinity Cloud," 1 Dec. 1945, ibid.; Hempelmann, "Itinerary of Trip Made by Colonel Warren, Captain Whipple and L. H. Hempelmann on 12 August 1945," 17 Aug. 1945, in Hoffman, "Health Physics Report," 77–80 (n. 44).

115. Hempelmann, "Itinerary," 79 (n. 114).

116. Hempelmann for files, "Trip to Ranches of Ted Coker and Mr. Raitliff on Sunday, 11th, November 1945," 14 Nov. 1945, in Hempelmann, "Biological Effects" (n. 114).

117. Hempelmann, "Itinerary," 79 (n. 114).

118. Hempelmann for files, 14 Nov. (n. 116); Hoffman, "Health Physics Report," 64 (n. 44).

119. Col. Stanley L. Stewart to Washington Headquarters, 6 Nov. 1945; Hempelmann for Trinity files, 1 Dec. (n. 114); Hempelmann for Trinity Follow-Up files, "Trip to Vicinity of Trinity to Purchase Cattle Alledgedly [sic] Damaged by the Atomic Bomb," 9 Dec. 1945.

120. Hempelmann for Trinity files, "Cattle Northeast of the Alamogordo Bombing Range Alledgedly [sic] Damaged by the Nuclear Explosion 16 July 1945," 1 April 1946, in Hempelmann, "Biological Effects" (n. 114); Hempelmann to Carroll L. Tyler, "Transfer of Alamogordo Cattle," 7 June 1948; Langham interview (n. 105); C. L. Comar, "The Fall-Out Problem," University of Tennessee–Atomic Energy Commission Agricultural Research Program, 1952, 1–2. Hempelmann, "Biological Effects" (n. 114) collects numerous additional observations, autopsies, and radiological tests on Trinity cattle.

121. Hempelmann for Trinity files, "Estimate of Dose of Beta Radiation Received by Cattle," 10 Feb. 1946.

122. Langham interview (n. 105).

123. J. F. Mullaney to Norris E. Bradbury, "Report by J. G. Hoffman on 'Biological Effects of July 16th Explosion,' " 3 Jan. 1946.

124. Weisskopf et al. to Kistiakowsky, 5 Sept. (n. 94).

125. Hirschfelder and Magee to Bainbridge, 14 Aug. (n. 104); Weisskopf et al. to Kistiakowsky, 5 Sept. (n. 94); Hoffman, "Health Physics Report," 3 (n. 44); Warren to Brig. Gen. James McCormack, Jr., 28 Nov. 1947.

126. Carl Buckland and A. Reinart to Hoffman, "Monitoring Trip Northeast of Trinity, 11 December 1945," 27 Dec. 1945, in Hoffman, "Health Physics Report," 82–85 (n. 44).

127. Warren to McCormack, 28 Nov. (n. 125); Szasz, *Day the Sun Rose Twice,* 136–141 (n. 15); Warren, "The 1948 Radiological and Biological Survey of Areas in New Mexico Affected by the First Atomic Bomb Test," UCLA Atomic Energy Project report no. UCLA-32, 17 Nov. 1949; Warren, "Role of Radiology," 885 (n. 21); Kermit H. Larson, "Continental Close-in Fallout: Its History, Measurement and Characteristics," in Vincent Schultz and Alfred W. Klement, Jr., eds., *Radioecology* (New York: Reinhold; Washington: American Institute of Biological Sciences, 1961), 19–21; Howard A. Hawthorne, ed., *Compilation of Local Fallout Data from Test Detonations, 1945–1962. Extracted from DASA 1251,* vol. 1: *Continental U.S. Tests,* Defense Nuclear Agency report no. DASA 1251(EX) (Washington, 1 May 1979), 4–7.

128. Aebersold, "Safety and Monitoring," 78 (n. 21).

129. Hempelmann for Trinity files, "Exposures of Military Personnel at Trinity (Since Aug. 14, 1945)," 29 Oct. 1945; Hempelmann to Bush, "Safety Instructions Concerning Military Personnel and Visitors in and around Crater Region," 22 Oct. 1945; Aebersold, "Safety and Monitoring," 78–89 (n. 21).

130. Groves, "Recollections," 27 (n. 1).

131. Hempelmann to Groves, 25 Aug. (n. 97). Cf. Hempelmann to Warren, 25 Aug. 1945.

132. Warren to Groves, 21 July (n. 45).

5. FROM JAPAN TO BIKINI

1. Col. Stafford L. Warren to Maj. Gen. Leslie R. Groves, "The Use of the Gadget as a Tactical Weapon Based on Observations Made During Test II," 25 July 1945; U.S. Strategic Bombing Survey, "Summary Report (Pacific War)," 1 July 1946, 15–17; David Hawkins, "Manhattan District History, Project Y, the Los Alamos Project," vol. 1: "Inception until August 1945," Los Alamos Scientific Laboratory [LASL] report no. LAMS-2532 (Vol. I), completed Aug. 1946, issued Dec. 1961, chap. 19; Wesley Frank Craven and James Lea Cate, eds., *The Army Air Forces in World War II,* vol. 5: *The Pacific: Matterhorn to Nagasaki, June 1944 to August 1945* (Chicago and London: University of Chicago Press, 1953), 716–717. The altitude at which both the Hiroshima and Nagasaki bombs

exploded was not measured and has since been much discussed; for an authoritative review, see *Hiroshima and Nagasaki: The Physical, Medical, and Social Effects of the Atomic Bombing,* ed. Committee for the Compilation of Materials on Damage Caused by the Atomic Bombs in Hiroshima and Nagasaki, trans. Eisei Ishikawa and David L. Swain (New York: Basic Books, 1981), 21–29.

2. Capt. William S. Parsons, USN, log of Hiroshima mission, 6 Aug. 1945, as published in Hawkins, "Project Y," 288 (n. 1). Cf. Fletcher Knebel and Charles W. Bailey II, *No High Ground* (New York: Harper & Row, 1960; repr. New York: Bantam Books, 1961), chaps. 8–10.

3. "Announced United States Nuclear Tests, July 1945 through December 1980," Office of Public Relations, DOE Nevada Operations Office, report no. NVO-209 (Rev. 1) (Las Vegas, Jan. 1981), 2.

4. Robert J. C. Butow, *Japan's Decision to Surrender* (Stanford: Stanford University Press; London: Geoffrey Cumberlege, Oxford University Press, 1954); Richard G. Hewlett and Oscar E. Anderson, Jr., *A History of the United States Atomic Energy Commission,* vol. 1: *The New World, 1939/1946* (University Park: Pennsylvania State University Press, 1962), 401–405; Barton J. Bernstein, "The Perils and Politics of Surrender: Ending the War with Japan and Avoiding the Third Atomic Bomb," *Pacific Historical Review* 46 (1977):1–28; Gregg Herken, *The Winning Weapon: The Atomic Bomb in the Cold War, 1945–1950* (New York: Alfred A. Knopf, 1980), chap. 1.

5. Robert S. Stone to Hymer L. Friedell, 9 Aug. 1945 (Chronological File of Miscellaneous Documents [Historical Interest], Argonne National Laboratory, Argonne, Ill.); Craven and Cate, eds., *Army Air Forces,* 5:722–723 (n. 1); Hewlett and Anderson, *New World,* 401 (n. 4).

6. *Hiroshima and Nagasaki,* 503–508 (n. 1).

7. Daniel Lang, "A Fine Moral Point," *New Yorker,* Jan. 1946, repr. in Lang, *From Hiroshima to the Moon* (New York: Dell, 1961), 52 (based on Lang's interview with Philip Morrison, one of the Tinian physicists who joined Farrell's group); Bernstein, "Perils and Politics," 3–8 (n. 4); Lisle A. Rose, *Dubious Victory: The United States and the End of World War II* (Kent, Ohio: Kent State University Press, 1973), 364; Michael J. Yavenditti, "The American People and the Use of Atomic Bombs on Japan: The 1940s," *Historian* 36 (1974):233–234; Paul Boyer, *By the Bomb's Early Light: American Thought and Culture at the Dawn of the Atomic Age* (New York: Pantheon Books, 1985), 187–188.

8. Brig. Gen. Thomas F. Farrell to Groves, "Report on Overseas Operations—Atomic Bomb," 27 Sept. 1945, as published in Anthony Cave Brown and Charles B. MacDonald, eds., *The Secret History of the Atomic Bomb* (New York: Dial Press/James Wade, 1977), 529, 534; "The Manhattan Project Atomic–Bomb Investigating Group," ibid., 552–582; Manhattan Project Atomic Bomb Investigating Group, "The Atomic Bombings of Hiroshima and Nagasaki," 30 June 1946, 1–2.

9. Army Chief of Staff Gen. George C. Marshall to Gen. Douglas MacArthur, Supreme Allied Commander in the Pacific, 12 Aug. 1945, as quoted in Investigating Group, "Atomic Bombings," 1–2 (n. 8).

10. Donald L. Collins, "Pictures from the Past: Journeys into Health Physics

in the Manhattan District and Other Diverse Places," in Ronald L. Kathren and Paul L. Ziemer, eds., *Health Physics: A Backward Glance. Thirteen Original Papers in the History of Radiation Protection* (New York: Pergamon Press, 1980), 40–41.

11. Investigating Group, "Atomic Bombings," 1 (n. 8); Stafford L. Warren, "The Role of Radiology in the Development of the Atomic Bomb," in Kenneth D. A. Allen, ed., *Radiology in World War II*, Medical Department in World War II: Clinical Series (Washington: Office of the Surgeon General, U.S. Army, 1966), 888, 890; interviews with Hymer L. Friedell, Cleveland, 12 June 1979; and Stafford L. Warren, Los Angeles, 30 Oct. 1979.

12. Col. Ashley W. Oughterson to Brig. Gen. Guy Denit, "Study of Casualty Producing Effects of Atomic Bombs," 28 Aug. 1945, as quoted by Oughterson, "Section 1: Introduction," in Oughterson, Shields Warren, Averill A. Liebow, George V. LeRoy, E. Cuyler Hammond, Henry L. Barnett, Jack D. Rosenbaum, and B. Aubrey Schneider, "Medical Report of the Joint Commission for the Investigation of the Effects of the Atomic Bomb in Japan," vol. 1, Sep. 1946, p. 7(1); James F. Nolan to Robert J. Buettner, 27 Oct. 1947; Joe W. Howland to Buettner, 5 Nov. 1947 (both Nolan and Howland recount the chronology of their trip to Japan with Warren; from the Stafford Warren Papers, University of California, Los Angeles); Oughterson and Shields Warren, eds., *Medical Effects of the Atomic Bomb in Japan*, National Nuclear Energy Series, Div. VIII, vol. 8 (New York: McGraw-Hill, 1956), 6–7, 431; Samuel Berg, "History of the First Survey on the Medical Effects of Radioactive Fall-Out," *Military Medicine* 124 (1959):782; Stafford Warren, "Role of Radiology," 888 (n. 11); interview with George V. LeRoy, Pentwater, Mich., 13 June 1979. Both Berg and LeRoy were members of Oughterson's team.

13. Oughterson, "Introduction," 3(1) (n. 12); Lang, "Fine Moral Point," 57 (n. 7); Nolan to Buettner, 27 Oct. (n. 12); Stafford Warren, "Role of Radiology," 888 (n. 11).

14. Farrell to Groves, 27 Sept., 535 (n. 8). Cf. Craven and Cate, eds., *Army Air Forces*, 721–722 (n. 1).

15. Lang, "Fine Moral Point," 57 (n. 7); Stafford Warren, "Role of Radiology," 888 (n. 11).

16. Farrell to Groves, 27 Sept., 536–537 (n. 8). Cf. Benedict R. Harris and Marvin A. Stevens, "Experiences at Nagasaki, Japan," *Connecticut State Medical Journal* 9 (1945):916–917; Craven and Cate, eds., *Army Air Forces*, 5:723–725 (n. 1).

17. Farrell to Groves, 27 Sept., 536, 537 (n. 8).

18. W. McRaney and J. McGahan, "Radiation Dose Reconstruction, U.S. Occupation Forces in Hiroshima and Nagasaki, Japan, 1945–1946," Defense Nuclear Agency [DNA] report no. 5512F (Washington, 6 Aug. 1980), 4–11, 15; "Hiroshima and Nagasaki Occupation Forces," DNA Public Affairs Office Fact Sheet (Washington, 6 Aug. 1980), 8–16; *Hiroshima and Nagasaki*, 73–79 (n. 1). All three works include citations to the literature on these much discussed subjects.

19. Warren, "Role of Radiology," 890 (n. 11). Cf. Collins, "Pictures from the Past," 41 (n. 10).

20. Investigating Group, "Atomic Bombings," 33 (n. 8).

21. Collins, "Pictures from the Past," 41 (n. 10).

22. Investigating Group, "Atomic Bombings," 27–34 (n. 8); Manhattan Project press release for 30 June 1946, "Manhattan Project Reports on Atom Bombings of Hiroshima and Nagasaki," 1 (in President's Secretary File, Papers of Harry S. Truman, Truman Library, Independence, Mo.; cited hereafter as Truman Papers, Sec. File); Joe W. Howland and Stafford L. Warren, "The Effects of the Atomic Bomb Irradiation on the Japanese," *Advances in Biological and Medical Physics* 1 (1948):388–392; "Hiroshima and Nagasaki Occupation Forces," 8 (n. 18).

23. Stafford Warren testimony in U.S. Congress, Senate, Special Committee on Atomic Energy, *Hearings*, 79 Cong., 1 and 2 Sess., 1945–46, 508–13, as cited in Yavenditti, "Use of Atomic Bomb," 234 (n. 7); Investigating Group, "Atomic Bombings," 28 (n. 8).

24. U.S. Strategic Bombing Survey, "The Effects of the Atomic Bombings of Hiroshima and Nagasaki," 19 June 1946, 16–17, 19–21. This summary report was intended for public distribution (Edwin A. Locke, Jr., to Pres. Truman, "Report of U.S. Strategic Bombing Survey," 21 June 1946 [Truman Papers, Sec. File]). The survey subsequently documented the bombings in full: *The Effects of the Atomic Bomb on Hiroshima, Japan* (3 vols.; Washington: Government Printing Office, 1947); and *The Effects of the Atomic Bomb on Nagasaki, Japan* (3 vols.; Washington: Government Printing Office, 1947). Cf. Craven and Cate, eds., *Army Air Forces*, 5:737–742 (n. 1).

25. Strategic Bombing Survey, "Effects," 16 (n. 24). Cf. Craven and Cate, eds., *Army Air Forces*, 5:722–725 (n. 1).

26. Frank Barnaby, "The Continuing Body Count at Hiroshima and Nagasaki," *Bulletin of the Atomic Scientists* 33 (Dec. 1977):48–53; "The Physical and Medical Effects of the Hiroshima and Nagasaki Bombs," report of the Natural Science Group organized by the International Peace Bureau, Geneva, ibid., 54–56; Peter Wyden, *Day One: Before Hiroshima and After* (New York: Simon & Schuster, 1984), book 2, "After the Bomb"; *Hiroshima and Nagasaki*, 364 (n. 1). Chaps. 3 and 10 of the last-cited work fully discuss the problem of determining bomb casualties and the various estimates made since 1945.

27. S. Okada, H. B. Hamilton, N. Egami, S. Okajima, W. J. Russell, and K. Takeshita, eds., "A Review of Thirty Years Study of Hiroshima and Nagasaki Atomic Bomb Survivors," *Journal of Radiation Research* 16 (1975), Supplement; John A. Auxier, *Ichiban: Radiation Dosimetry for the Survivors of Hiroshima and Nagasaki* (Oak Ridge, Tenn.: Technical Information Center, 1977); W. E. Loewe and E. Mendelsohn, "Revised Dose Estimates at Hiroshima and Nagasaki," *Health Physics* 41 (1981):663–666; T. Straume and R. Lowry Dobson, "Implications of New Hiroshima and Nagasaki Dose Estimates: Cancer Risks and Neutron RBE," *Health Physics* 41 (1981):666–671; Victor P. Bond and J. W. Thiessen, eds., *Reevaluation of Dosimetric Factors: Hiroshima and Nagasaki*, Report no. CONF-810928 (Oak Ridge, Tenn.: Technical Information Center, 1982). John W. Gofman, *Radiation and Human Health* (San Francisco: Sierra Club Books, 1981), passim, argues that doses assigned to Japanese survivors are too low. He per-

suades few experts, who refute his assumptions, methods, and results in this matter, as in his broader challenge to accepted radiation safety standards; see, e.g., Jane M. Orient, "A Critique of the Statistical Methodology in *Radiation and Human Health* by John W. Gofman," *Health Physics* 45 (1983):823–827.

28. "Nuclear Test Personnel Review (NTPR)," DNA Public Affairs Office Fact Sheet (Washington, 1 March 1982); McRaney and McGahan, "Radiation Dose Reconstruction" (n. 18); "Hiroshima and Nagasaki Occupation Forces" (n. 18). Cf. National Research Council, "Report of Panel on Feasibility and Desirability of Performing Epidemiological Studies on U.S. Veterans of Hiroshima and Nagasaki," 21 Aug. 1981; National Research Council, "Multiple Myeloma among Hiroshima/Nagasaki Veterans," 15 July 1983.

29. Office of Technology Assessment, Health Program, "Review of the Report, 'Multiple Myeloma among Hiroshima/Nagasaki Veterans,' " Staff Memorandum, 17 Dec. 1983; R. Jeffrey Smith, "Study of Atomic Veterans Fuels Controversy," *Science* 221 (1983):733–734. For a fully documented journalistic critique of the DNA findings, see Harvey Wasserman and Norman Solomon, with Robert Alvarez and Eleanor Walters, *Killing Our Own: The Disaster of America's Experience with Atomic Radiation* (New York: Dell, 1982), chap. 1.

30. Col. H. W. Allen (Asst. Adjutant General, Office of the Supreme Commander for the Allied Powers) to Commanding Officer, Eighth Army, "Atomic Bomb Investigation," 12 Oct. 1945, as quoted in Oughterson and Shields Warren, eds., *Medical Effects*, 9 (n. 12); *Hiroshima and Nagasaki*, 505, 508–509 (n. 1).

31. Masao Tsuzuki, "Experimental Studies on the Biological Action of Hard Roentgen Rays," *American Journal of Roentgenology* 16 (1926):134–150, as cited in A. Lundie, "Some Medical Aspects of Atomic Warfare," *Journal of the Royal Army Medical Corps* 94 (1950):246.

32. Morrison as quoted in Lang, "Fine Moral Point," 54–55 (n. 7).

33. Oughterson and Shields Warren, eds., *Medical Effects*, 7–8, 435–443 (n. 12); *Hiroshima and Nagasaki*, 509 (n. 1); Berg, "History of the First Survey," 782–785 (n. 12); Shields Warren, "Hiroshima and Nagasaki Thirty Years After," *Proceedings of the American Philosophical Society* 121 (April 1977):97–99.

34. Nello Pace and Robert E. Smith, "Measurement of the Residual Radiation Intensity at the Hiroshima and Nagasaki Atomic Bomb Sites," Naval Medical Research Institute report no. 160A, 16 April 1946; Oughterson and Shields Warren, eds., *Medical Effects*, 439–441 (n. 12); "Hiroshima and Nagasaki Occupation Forces," 8 (n. 18).

35. Berg, "History of the First Survey," 783 (n. 12).

36. Shunzo Okajima, Kenji Takeshita, Shigetoshi Antoku, Toshio Shiomi, Walter J. Russell, Shoichiro Fujita, Haruma Yoshinaga, Shotaro Neriishi, Sadahisa Kawamoto, and Toshiyuki Norimura, "Radioactive Fallout Effects of the Nagasaki Atomic Bomb," *Health Physics* 34 (1978):621–633.

37. Oughterson and Shields Warren, eds., *Medical Effects* (n. 12); Oughterson et al., "Medical Report" (n. 12); Oughterson and Shields Warren, eds., *Medical Effects of Atomic Bombs: The Report of the Joint Commission for the Effects of the Atomic Bomb in Japan* (6 vols.; Washington: Atomic Energy Commission [AEC], 1951).

38. Science Council of Japan, ed., *Collection of the Reports on the Investigation of the Atomic Bomb Casualties* [in Japanese] (Tokyo, 1951), and idem, same title (2 vols.; Tokyo, 1953), as cited in *Hiroshima and Nagasaki,* 507–509 (n. 1); Auxier, *Ichiban,* 4 (n. 27). The Stafford Warren Papers include English translations of many of the original Japanese reports.

39. Shields Warren, "The Physiological Effects of Radiant Energy," *Annual Review of Physiology* 7 (1945):61–74, reviews prewar knowledge.

40. E.g., William L. Laurence, *Dawn over Zero: The Story of the Atomic Bomb* (2d ed.; New York: Alfred A. Knopf, 1946; repr. Westport, Conn.: Greenwood Press, 1972), 244–252; John Hersey, *Hiroshima* (New York: Alfred A. Knopf, 1946).

41. See, e.g., Shields Warren, "The Pathologic Effects of an Instantaneous Dose of Radiation," *Cancer Research* 6 (1946):449–453; Shields Warren and R. H. Draeger, "The Pattern of Injuries Produced by the Atomic Bombs at Hiroshima and Nagasaki," *U.S. Naval Medical Bulletin* 46 (1946):1349–1353; Charles L. Dunham, "Symposium on Medicine and Atomic Warfare: Medical Aspects of Atomic Warfare," *Connecticut State Medical Journal* 15 (1951):1039–1047; Dunham, Eugene P. Cronkite, George V. LeRoy, and Shields Warren, "Atomic Bomb Injury: Radiation," *Journal of the American Medical Association* 147 (1951):50–54. Other early reports are cited in *Hiroshima and Nagasaki,* 509 (n. 1).

42. *Hiroshima and Nagasaki,* chap. 8 (n. 1), summarizes the acute effects of the bombings.

43. National Research Council, Committee on Atomic Casualties, Genetics Conference, "Genetic Effects of Atomic Bombs in Hiroshima and Nagasaki," *Science* 106 (1947):331–333; AEC, *Atomic Energy and the Life Sciences* (Washington: Government Printing Office, 1949), 35–46; G. W. Beadle, D. R. Charles, C. C. Craig, L. H. Snyder, and Curt Stern, "Statement of the Conference in Genetics to the Committee on Atomic Casualties," 11 July 1953, App. B to James V. Neel, "Genetics Conference," University of Michigan, 10–11 July 1953. Cf. William J. Schull, Masanori Otake, and James V. Neel, "Genetic Effects of the Atomic Bomb: A Reappraisal," *Science* 213 (1981):1220–1227.

44. Secretary of the Navy James Forrestal to President Harry S. Truman, 18 Nov. 1946, proposing a long-term study, with Truman's approval dated 26 Nov. 1946.

45. *Atomic Energy and the Life Sciences,* 47–55 (n. 43); Richard G. Hewlett and Francis Duncan, *A History of the United States Atomic Energy Commission,* vol. 2: *Atomic Shield, 1947/1952* (University Park: Pennsylvania State University Press, 1969; repr. Washington: AEC, 1972), 244–245; Yavenditti, "Use of Atomic Bomb," 234 and passim (n. 7). Cf. Hiroshi Maki, Isamu Nagai, Atsumu Okada, Mitsugu Usagawa, and Kimiko Ono, *The Atomic Bomb Casualty Commission, 1947–1975: A General Report on the ABCC-JNIH Joint Research Program,* report no. TID-28719, National Academy of Sciences–National Research Council and Japanese National Institute of Health of the Ministry of Health and Welfare, Washington and Tokyo, 1978; Hymer L. Friedell, "Early Atomic Bomb Casualty Commission Perceptions and Planning," in Bond and Thiessen, eds., *Reevaluations*

of *Dosimetric Factors*, 1–5 (n. 27). On the role of disaster studies in furthering knowledge, see William W. Lowrance, *Of Acceptable Risk: Science and the Determination of Safety* (Los Altos, Cal.: William Kaufmann, 1976), 49–51.

46. Walter Millis, *Arms and Men: A Study in American Military History* (New York: G. P. Putnam's Sons, 1956; repr. New York: Mentor Books, n.d.), 224-232, 277; Hewlett and Anderson, *New World*, 581–582 (n. 4); Kenneth L. Moll, "Operation Crossroads: The Bikini A-Bomb Tests, 1946," *Air Force Magazine* 54 (July 1971):62–63; Lloyd J. Graybar, "Bikini Revisited," *Military Affairs* 44 (1980):118–119; Herken, *Winning Weapon*, 224–225 (n. 4).

47. Vice Adm. E. L. Cochrane (Chief, Bureau of Ships) and Rear Adm. G. F. Hussey, Jr. (Chief, Bureau of Ordnance), to Chief of Naval Operations [CNO], 1 Oct. 1945, as quoted in William A. Shurcliff (Joint Task Force One [JTF-1] Historian), "Technical Report of Operation Crossroads," Crossroads report no. XRD-208, 18 Nov. 1946, 1.4. Cf. Rear Adm. T. A. Solberg (Director of Ship Material [DSM], JTF-1), "Operation Crossroads General Information Bulletin," n.d., 4 (Stafford Warren Papers).

48. Commodore William S. Parsons to Norris E. Bradbury (LASL Director), "Possible Tests of Atomic Bombs against Naval Vessels," 26 Oct. 1945 (in Mail and Records Center, Los Alamos Scientific Laboratory, Los Alamos, N.M.; hereafter cited as LASL Records); Commander, Joint Task Force One [CJTF-1], "Report on Atomic Bomb Tests Able and Baker (Operation Crossroads) Conducted at Bikini Atoll, Marshall Islands, on 1 July 1946 and 25 July 1946," 15 Nov. 1946, pp. I-(B)-1, -2; Shurcliff, "Technical Report," 1.3–5 (n. 47); Shurcliff, *Bombs at Bikini: The Official Report of Operation Crossroads* (New York: Wm. H. Wise, 1947), 10–11.

49. CJTF-1, "Report," I-(B)-2, -3 (n. 48); Shurcliff, "Technical Report," 1.5–6 (n. 47); Shurcliff, *Bombs at Bikini*, 11–15 (n. 48); Moll, "Operation Crossroads," 63–64 (n. 46); Neal O. Hines, *Proving Ground: An Account of the Radiobiological Studies in the Pacific, 1946–1961* (Seattle: University of Washington Press, 1962), 21–22; L. Berkhouse, S. E. Davis, F. R. Gladeck, J. H. Hallowell, C. B. Jones, E. J. Martin, F. W. McMullan, and M. J. Osborne, *Operation Crossroads, 1946*, report no. DNA 6032F (Washington, 1 May 1984), chap. 1.

50. Lt. Cmdr. W. M. Rigdon to Matthew Connelly, 3 April 1946, w/att. "Purposes of Atomic Bomb Tests and Reasons for Conducting Them at an Early Date" (Truman Papers, Sec. Files); Herbert Scoville, Jr. (for Stafford Warren), to Ralph A. Sawyer (Technical Director, JTF-1), "Radioactivity in Target Area, Measurement of," 11 April 1946; "Ship Preparation Plan," Annex W to CJTF-1, Operation Plan no. 1–46, 15 April 1946; "BuMed Research Section Inspections," App. III to "Reboarding and Inspection Plan," Annex X to CJTF-1, Operation Plan no. 1–46; CJTF-1, "Report," I-(B)-2, -3 (n. 48); Shurcliff, "Technical Report," chap. 2, "Objects of the Tests" (n. 47); Office of the Historian, JTF-1, *Operation Crossroads: The Official Pictorial Record* (New York: Wm. H. Wise, 1946), 67, 108; Shurcliff, *Bombs at Bikini*, 2–5 (n. 48); R. H. Draeger and Shields Warren, "Medicine at the Crossroads," *U.S. Naval Medical Bulletin* 47 (1947):219–225; Hines, *Proving Ground*, 35 (n. 49); Berkhouse et al., *Operation Crossroads*, chap. 3, "Crossroads Experimental Program" (n. 49).

51. Brig. Gen. A. J. McFarland (Secretary, Joint Chiefs of Staff [JCS]) to W. H. P. Blandy, 11 Jan. 1946, as quoted in Shurcliff, *Bombs at Bikini*, 14–15 (n. 48). This directive was issued to Blandy by the joint chiefs on his assignment as CJTF-1.

52. CJTF-1, "Report," I-(B)-3, -4, -5 (n. 48); Shurcliff, *Bombs at Bikini*, 12, 17, 21, 27 (n. 48); Hines, *Proving Ground*, 22 (n. 49); Graybar, "Bikini Revisited," 119 (n. 46).

53. As quoted in "Operation Crossroads," *National Geographic* 91 (April 1947), caption for plate XI.

54. CJTF-1, Operation Plan no. 1–46, 21 Jan. 1946 (revised and reissued 15 April 1946 after the president postponed the tests from May to July); Blandy to CNO et al., "General Information on Atomic Bomb Tests," 26 Jan. 1946; Headquarters Army Service Forces to Commanding General, San Francisco Port of Embarkation, et al., "CROSSROADS Project—Atomic Bomb Tests—Pacific," 6 Feb. 1946 (LASL Records).

55. "Preliminary Planning," in CJTF-1, "Report," V-(A)-3 (n. 48); Shurcliff, *Bombs at Bikini*, 16–17 (n. 48); Hines, *Proving Ground*, 22–24 (n. 49).

56. Stafford Warren and Capt. George M. Lyon to CJTF-1, "Evacuation of Ronjerik [sic]," 6 March 1946; Stafford Warren to CJTF-1, "Evacuation of Rongerik and Rongelap," 8 March 1946; Stafford Warren to J-3 (Capt. C. H. Lyman), "Evacuation of Atolls Neighboring to Bikini," 13 March 1946; Carl Markwith, "Farewell to Bikini," *National Geographic* 90 (July 1946):97–116; Hines, *Proving Ground*, 24–25 (n. 49); Merze Tate and Doris M. Hull, "Effects of Nuclear Explosions on Pacific Islanders," *Pacific Historical Review* 33 (1964):380–381; Robert C. Kiste, *The Bikinians: A Study in Forced Migration* (Menlo Park, Cal.: Cummings, 1974), chap. 2; Jonathan M. Weisgall, "The Nuclear Nomads of Bikini," *Foreign Policy* no. 39 (Summer 1980):75–80.

57. "Preliminary Planning," V-(A)-2, -3 (n. 55); "Advance Preparations; Training of Task Groups," in CJTF-1, "Report," V-(B)-4 (n. 48); Hines, *Proving Ground*, 33–34 (n. 49).

58. Bradbury to All Division Leaders, "B-Division," 24 Jan. 1946 (LASL Records); "Staff Organization—Deputy Task Force Commander for Technical Direction," 1 March 1946, reproduced in CJTF-1, "Report," IV-(B)-fig. 4 (n. 48); "History of Los Alamos B Division," 30 April 1946 (LASL Records); Shurcliff, "Technical Report," 3.24–5, 3.34 (n. 47); CJTF-1, "Report," I-(B)-6 (n. 48); Shurcliff, *Bombs at Bikini*, 46–47 (n. 48).

59. Col. H. C. Gee to Bradbury, "Crossroads Summary no. 8," 31 Jan. 1946. Cf. "Los Alamos B Division Measurement Section Technical Responsibilities," 6 May 1946 (LASL Records).

60. Lyon (for Stafford Warren), "Safety Plan (Tentative)," draft, 18 Jan. 1946; Lyon, "Medical Service Plan (Tentative)," 18 Jan. 1946; "Special Reports: Safety Plan (Including Radiological)," in CJTF-1, "Report," VII-(C)-1, -2 (n. 48); Stafford Warren, "Role of Radiology," 903 (n. 11).

61. "Summation by Commander Joint Task Force One," in CJTF-1, "Report," II-(A)-2 (n. 48).

62. Stafford Warren to Bradbury, 22 Jan. 1946 (LASL Records); Bradbury to

Stafford Warren, 23 Jan. 1946 (LASL Records); Scoville (for Stafford Warren) to Sawyer, "Organization of Radiological Safety Section [RSS]," 25 April 1946 (Stafford Warren Papers); Organization chart, "The Radiological Safety Section," n.d. (Stafford Warren Papers); Shurcliff, "Technical Report," 3.42, 3.57 (n. 47); Scoville, "The Atomic Bomb and the Resultant Phenomena," in *Radiological Defense*, vol. 3: *A Series of Indoctrination Lectures on Atomic Explosion, with Medical Aspects*, ed. Armed Forces Special Weapons Project (Washington, n.d.), 16–22; James P. Cooney, "Medical Effects of Atomic Explosion," ibid., 30–36.

63. Robert R. Newell "Report of Medic[o]-Legal Board," 19 Aug. 1946 (Stafford Warren Papers).

64. Shurcliff, "Technical Report," 3.48–49, 3.52–54, 3.56 (n. 47); George M. Lyon, "Radiological Defense," in *Radiological Defense*, 3:97–109 (n. 62).

65. "Radiological Safety Plan Test Able," App. II to "Safety Plan," Annex E to CJTF-1, Operation Plan no. 1–46, E-II-1 (n. 50).

66. "Safety Plan," E-1 (n. 65).

67. "General Considerations of Radiological Safety Test Able," App. I to "Safety Plan," E-I-1 to E-I-3 (n. 65); John L. Magee to Bradbury, "Protection from Gamma Radiation," 31 May 1946 (LASL Records); Scoville, "Estimation of Hazard from Airborne Radioactive Material," May 1946 (LASL Records).

68. Henry W. Newson to Bradbury, "Possible Difficulties in Naval Tests," 17 Dec. 1945, 4 (LASL Records).

69. Blandy to CNO et al., 26 Jan., 7 (n. 54).

70. "Re-Entry Plan for Test Baker," draft App. II to "Re-Entry Plan," Annex I to CJTF-1, Operation Plan no. 1-46 (n. 50).

71. "Safety Prediction—Test Baker," draft, May 1946, 2–3 (Stafford Warren Papers).

72. "General Considerations of Radiological Safety," E-I-2 (n. 67). Emphasis in original.

73. "Radiological Safety Plan Test Able," E-II-1 to -8 (n. 65); "Logistics," App. VII to "Safety Plan," E-VII-2 (n. 65). Cf. Berkhouse et al., *Operation Crossroads*, chap. 2, "Radiological Safety" (n. 49).

74. "Radiological Safety Plan Test Able," E-II-1 (n. 65).

75. Ibid., E-II-8, -9; Gerhard Dessauer (Photometry Group Leader), "Instructions to Senior Monitors (Airborne or Waterborne)," for Able and Baker, n.d.; Dessauer, "Personnel Film Badges [Able]," 22 July 1946; Dessauer, "Photographic Dosimetry," excerpts from a lecture, Aug. 1947, p. 4. The preceding documents are in "Continent and Pacific Historical Records," Box 3, REECo microfilm records, roll 1.

76. Dessauer, "Personnel Film Badges" (n. 75); Dessauer, "Photographic Dosimetry," 3–4 (n. 75); "Logistics," E-VII-2 (n. 73).

77. "General Considerations of Radiological Safety," E-I-3 (n. 67).

78. Lyon to Parsons, "Conference on Radiological Measurements," 30 Jan. 1946 (LASL Records).

79. "General Considerations of Radiological Safety," E-1-3 (n. 67).

80. "Radiological Safety Plan Test Able," E-II-9 (n. 65).

81. Sawyer to Parsons, "Instrumentation for Crossroads Project," 8 Feb. 1946

(LASL Records); Birchard M. Brundage to All Concerned, "Itinerary of Training Group," 18 Feb. 1946, w/attachments (LASL Records); "Special Reports: Safety," VII-(C)-2, -4, -6 (n. 60); Shurcliff, *Bombs at Bikini*, 31 (n. 48).

82. "Outline [of RadSafe Plan]," 4 Feb. 1946; Stafford Warren, "Radiological Safety," in Sawyer to Parsons, 8 Feb., 12 (n. 81); "Radiological Safety Annex to Safety Plan—Operation Crossroads," draft, 9 Feb. 1946; Lt. Col. John M. Talbot, "Annex_____to Air Ops. Order No. 1," draft, 22 Feb. 1946; Talbot to Stafford Warren, 23 Feb. 1946.

83. Lyon (for Stafford Warren) to Cmdr. F. L. Ashworth (Parson's Assistant for Aviation), "Crossroads Operation—Time Needed between Shots A and B," 6 Feb. 1946; "Special Reports: Safety," VII-(C)-2, -4 (n. 60).

84. Lyon to Ashworth, 6 Feb. (n. 83). Emphasis in original.

85. Hewlett and Anderson, *New World*, 582 (n. 4); Herken, *Winning Weapon*, 175–176 (n. 4); Lloyd J. Graybar and Ruth Flint Graybar, "America Faces the Atomic Age, 1946," *Air University Review* 35 (Jan.-Feb. 1984):69–70.

86. CJTF-1, "Report," I-(B)-11, -12 (n. 48); "Preliminary Planning," V-(A)-5, -6 (n. 55); Moll, "Operation Crossroads," 64 (n. 46); Graybar, "Bikini Revisited," 119 (n. 46).

87. Henry A. Wallace's notes on the cabinet meeting of 19 March, three days before the publicly announced postponement, as quoted in Herken, *Winning Weapon*, 176 (n. 4). Cf. Graybar and Graybar, "America Faces the Atomic Age," 71–72 (n. 85).

88. Commander, Western Sea Frontier [CWSF], to CNO, 1 Feb. 1946 (in "Crossroads Dispatches" folder, Crossroads box 47, DASIAC Library, Kaman Tempo, Santa Barbara, Cal.; hereafter cited as Crossroads Dispatches).

89. ComServPac to various commands, 9 April 1946; CNO to various commands, 9 April 1946; ComServPac to various commands, 19 April 1946 (Crossroads Dispatches). See also "Advance Preparations," V-(B)-5 (n. 57); "Summation by Commander," II-(A)-3 (n. 61); CJTF-1, "Report," I-(B)-9 (n. 48); "Comments by the Assistant Chief of Staff for Personnel," ibid., III-(A)-3.

90. Shurcliff, *Bombs at Bikini*, 20–22 (n. 48).

91. Lyon to Parsons, "Personnel for Radiological Safety Section," 4 April 1946 (Stafford Warren Papers).

92. "Special Reports: Safety," VII-(C)-5 (n. 60).

93. Ibid., VII-(C)-5, -6.

94. "Personnel Aboard U.S.S. Haven (AH-12)," 3 June 1946 (Stafford Warren Papers); "Special Reports: Safety," VII-(C)-4 (n. 60).

95. Stafford Warren to James Nolan, 19 April 1946 (LASL Records; similar letters went to all members of the rad-safe section); "Special Reports: Safety," VII-(C)-5, -7, -8 (n. 60); Stafford Warren, "Role of Radiology," 904–908 (n. 11).

96. "Climatology April through October," App. I to "Aerological Plan," Annex T to CJTF-1, Operation Plan no. 1–46, T-I-1 (n. 50); "Preliminary Planning," V-(A)-6 (n. 55); CJTF-1, "Report," I-(B)-13 (n. 48); "Special Reports: Air Operations," ibid., VII-(E)-59, -60; Shurcliff, *Bombs at Bikini*, 102 (n. 48); Hines, *Proving Ground*, 33 (n. 49); Moll, "Operation Crossroads," 64 (n. 46).

97. A. A. Cumberledge, "Aerological Aspects of the Bikini Bomb Tests," *Scientific Monthly* 64 (Feb. 1947):136–137.

98. As quoted in ibid., 138.

99. "Climatology April through October," T-2 (n. 96); Shurcliff, "Technical Report," 3.32–33 (n. 47); "Conduct of the Tests from the Operational Aspect," in CJTF-1, "Report," V-(C)-2 (n. 48); "Air Operations," VII-(E)-13, -39, -60 to -63 (n. 96); Cumberledge, "Aerological Aspects," 140–142 (n. 97).

100. CJTF-1, "Report," I-(B)-13 (n. 48); "Conduct of the Tests," V-(C)-2 (n. 99); "Air Operations," VII-(E)-59, -60, -62 (n. 96); Cumberledge, "Aerological Aspects," 138–140, 142 (n. 97).

101. RSS Briefing Sheet, "Schedule [Queen day plan]," 22 June 1946 (Stafford Warren Papers); "Chronology of QUEEN DAY Operation," in CJTF-1, "Report," VI-(A)-1, -2, (n. 48); CJTF-1, "Aerological Report on Operation Crossroads," OPNAV-JTF-P1001, n.d., 15; Cumberledge, "Aerological Aspects," 142 (n. 97); David Bradley, *No Place to Hide* (Boston: Little, Brown, 1948), 32. Bradley was one of the original group of army doctors assigned to Crossroads radiological safety in Jan. 1946. Cast in the form of his personal log, the book actually draws on his frequent and lengthy letters home; see Alexander Hammond's review of the recently reissued book in *Bulletin of the Atomic Scientists* 39 (Nov. 1983):37–39.

102. "Chronology of QUEEN DAY Operation," VI-(A)-2, -3, -4, -5 (n. 101); "Air Operations," VII-(E)-116 (n. 96); Bradley, *No Place to Hide*, 39 (n. 101).

103. Parsons to JTF-1 Asst. Chief of Staff for Operations, J3 (Capt. C. H. Lyman) et al., "Drone Boat Operation," 18 April 1946 (LASL Records); Solberg to Parsons, "[DSM] Situation Report no. 17," 25 June 1946 (LASL Records); "Chronology of QUEEN DAY Operation," VI-(A)-5, -6, -7 (n. 101); Bradley, *No Place to Hide*, 37–40 (n. 101).

6. CROSSROADS

1. David Bradley, *No Place to Hide* (Boston: Little, Brown, 1948), 46.

2. Commander, Joint Task Force One (CJTF-1), "Aerological Report on Operation Crossroads," OPNAV-JTF-P1001, n.d., 18; A. A. Cumberledge, "Aerological Aspects of the Bikini Bomb Tests," *Scientific Monthly* 64 (Feb. 1947):143.

3. CJTF-1 to Joint Task Force One (JTF-1), "Preparation for Tactical Contingencies on Able Day," 29 June 1946 (in Mail and Records Center, Los Alamos Scientific Laboratory, Los Alamos, N.M.; hereafter cited as LASL Records).

4. As quoted in Cumberledge, "Aerological Aspects," 143 (n. 2).

5. Col. Stafford L. Warren to JTF-1 Asst. Chief of Staff for Operations, "Selection of 1 July as Possible Able Day, Reasons for," 30 June 1946 (Stafford Warren Papers, University of California, Los Angeles); "Chronology of ABLE DAY Operation," in CJTF-1, "Report on Atomic Bomb Tests Able and Baker (Operation Crossroads) Conducted at Bikini Atoll, Marshall Islands, on 1 July 1946 and 25 July 1946," 15 Nov. 1946, VI-(B)-1.

6. Bradley, *No Place to Hide*, 46–47 (n. 1).

7. "Chronology of ABLE DAY Operation," VI-(B)-2 to -5 (n. 5); Cumberledge, "Aerological Aspects," 143 (n. 2).

8. "Chronology of ABLE DAY Operation," VI-(B)-3, -4 (n. 5); "Special Reports: Air Operations," in CJTF-1, "Report," VII-(E)-132 to -135 (n. 5); Cumberledge, "Aerological Aspects," 141, 143–144 (n. 2).

9. Marshall G. Holloway to Ralph A. Sawyer, "Annex H, Bikini Evacuation Plan," 11 March 1946 (LASL Records); J. H. Nolan to Stafford Warren, "Monitoring Problems on Kwajalein," 11 June 1946 (Stafford Warren Papers); "Chronology of ABLE DAY Operation," VI-(B)-5, -6 (n. 2); "Air Operations," VII-(E)-149 to -151, -164, -173 (n. 8); William A. Shurcliff, "Technical Report of Operation Crossroads," Crossroads report no. XRD-208, 18 Nov. 1946, 10.3–4; Shurcliff, *Bombs at Bikini: The Official Report of Operation Crossroads* (New York: Wm. H. Wise, 1947), 104.

10. "Air Operations Requiring Instruments or Monitors," Feb. 1946; unsigned memo to Stafford Warren, "Test Able Only, Radiological Air Data, Instruments/ Monitors," 28 June 1946 (both in Stafford Warren Papers); Gerhard Dessauer, "Instructions to Senior Monitors (Airborne or Waterborne)," for Able and Baker, n.d. (in "Continent and Pacific Historical Records," Box 3, Reynolds Electrical & Engineering Co., Inc. [REECo], microfilm records, roll 1); Bradley, *No Place to Hide*, 50–51 (n. 1).

11. Bradley, *No Place to Hide*, 51 (n. 1).

12. "Chronology of ABLE DAY Operation," VI-(B)-6 (n. 5); "Air Operations," VII-(E)-151 (n. 8).

13. Bradley, *No Place to Hide*, 53 (n. 1).

14. "Chronology of ABLE DAY Operation," VI-(B)-7 (n. 5); "Air Operations," VII-(E)-151, -152 (n. 8).

15. Bradley, *No Place to Hide*, 54 (n. 1). Cf. "Air Communications Plan," App. V to "Air Plan," Annex F to CJTF-1, Operation Plan no. 1–46, 15 April 1946, F-V-7.

16. R. S. Warner to E. B. Doll, "Proposed Schedule for CROSSROADS Training Program," 25 Feb. 1946 (LASL Records); Radiological Safety Section [RSS], "Practice Involving Pilots Participating in Bikini Tests," n.d. (Stafford Warren Papers); "Air Operations," VII-(E)-153 (n. 8); Shurcliff, *Bombs at Bikini*, 105 (n. 9).

17. "Chronology of ABLE DAY Operation," VI-(B)-7 (n. 5); "Air Operations," VII-(E)-153 (n. 8).

18. Bradley, *No Place to Hide*, 54–55 (n. 1).

19. Shurcliff, *Bombs at Bikini*, 36 (n. 9); Kenneth L. Moll, "Operation Crossroads: The Bikini A-Bomb Tests, 1946," *Air Force Magazine* 54 (July 1971):65, 67.

20. S. D. Kirkpatrick, diary for 1–28 July 1946, as published in "A-Bomb Tests: As Viewed by the Editor," *Chemical Engineering* 53 (Aug. 1946):94–96ff., entry for 1 July.

21. William L. Laurence, *Dawn over Zero: The Story of the Atomic Bomb* (2d ed.; New York: Alfred A. Knopf, 1946; repr. Westport, Conn.: Greenwood Press, 1972), 285. Cf. President's Evaluation Commission, "Preliminary Report on

First Bikini Atom Bomb Test," released 11 July 1946; Lloyd J. Graybar and Ruth Flint Graybar, "America Faces the Atomic Age, 1946," *Air University Review* 35 (Jan.-Feb. 1984):72–73; Paul Boyer, *By the Bomb's Early Light: American Thought and Culture at the Dawn of the Atomic Age* (New York: Pantheon Books, 1985), chap. 8, "The Mixed Message of Bikini," 82–92. For evidence of changing public reaction, see Sylvia Eberhart, "How the American People Feel about the Atomic Bomb," *Bulletin of the Atomic Scientists* 3 (June 1947).146–149f.; Leonard S. Cottrell, Jr., and Eberhart, *American Opinion on World Affairs in the Atomic Age* (Princeton: Princeton University Press, 1948); Hazel Gaudet Erskine, "The Polls: Atomic Weapons and Nuclear Energy," *Public Opinion Quarterly* 27 (1963):155–190.

22. "Air Radiological Safety Plan," App. XII to "Air Plan" (n. 15); "Radiological Danger Areas for Air Operations," 27 June 1946 (Stafford Warren Papers); "Chronology of ABLE DAY Operation," VI-(B)-7 (n. 5); "Air Operations," VII-(E)-139, -140, -157, -158 (n. 8); "Special Reports: Safety Plan (Including Radiological)," in CJTF-1, "Report," VII-(C)-9, -12 (n. 5).

23. Capt. George M. Lyon to JTF-1 Deputy Commander for Aviation, "Pilot and Monitor Training over 'Hot' Area," 28 Feb. 1946; John W. Talbot and James F. Nolan to CJTF-1, "Pilot and Monitor Training over 'Hot' Area," 9 March 1946 (both in LASL Records); "Special Reports: Safety," VII-(C)-10, -11 (n. 22).

24. "Special Reports: Safety," VII-(C)-10 (n. 22).

25. CJTF-1 to CincPac, 091830Z, 9 March 1946 (LASL Records); "Chronology of ABLE DAY Operation," VI-(B)-8 (n. 5); "Air Evacuation Plan," App. VIII to "Air Plan," F-VIII-1 (n. 15); "Special Reports: Ship Operations," in CJTF-1, "Report," VII-(F)-24, -33, -34 (n. 5); Cumberledge, "Aerological Aspects," 146–147 (n. 2). Note that all untitled dispatches, as CJTF-1 to CincPac, above, are identified by a 6-digit code: the first 2 digits are the day of the month, the next 4 the time of day on the 24-hour military clock; the concluding "Z" indicates military Zulu (i.e., Greenwich) time. In the text, I have used local (Bikini) times, which are 11 hours later.

26. Stafford Warren to JTF-1 Asst. Chief of Staff for Operations, "B 29's for Safety Reconnaissance, Request for," 13 March 1946 (Stafford Warren Papers); Radiological Safety Advisor (Stafford Warren) to Deputy Commander for Technical Direction (DepComTech), 211445Z, 21 June 1946 (LASL Records); unsigned notes, "Resume of Able Day Operations—Col. Warren," 6 July 1946, 2–3 (Stafford Warren Papers); "Chronology of ABLE DAY Operation," VI-(B)-13 (n. 5); "Air Operations," VII-(E)-146, -147, -160 to -162 (n. 8); "Special Reports: Safety," VII-(C)-10 (n. 22); Bradley, *No Place to Hide*, 62–63 (n. 1).

27. "Chronology of ABLE DAY Operation," VI-(B)-8, -9 (n. 5); Bradley, *No Place to Hide*, 49–50 (n. 1).

28. Bradley, *No Place to Hide*, 58 (n. 1).

29. Joint Chiefs of Staff [JCS] Evaluation Board for Atomic Bomb Tests, "Preliminary Report Following the First Atomic Bomb Test," 11 July 1946; JTF-1 Director of Ship Material [DSM], "Technical Inspection Report: Test A and Test B," Bureau of Ships [BuShips] Group Final Report XRD-2, vol. 1, by Capt. F. X. Forest, 16 Dec. 1946, 4–27; Moll "Operation Crossroads," 64 (n. 19).

30. CJTF-1 to JCS, 010206Z, 1 July 1946, as quoted in "Chronology of ABLE DAY Operation," VI-(B)-11 (n. 5).

31. Bradley, *No Place to Hide*, 58 (n. 1).

32. Paul W. Tibbets, with Clair Stebbins and Harry Franken, *The Tibbets Story* (New York: Stein & Day, 1978), chap. 33.

33. Gen. James Doolittle as quoted in Moll, "Operation Crossroads," 69 (n. 19).

34. President's Evaluation Commission, "Preliminary Report [Able]" (n. 21); JCS Evaluation Board, "Preliminary Report Following the Second Atomic Bomb Test," 30 July 1946, Sect. I: "Supplement to Preliminary Report on Test 'A' "; "Air Operations," VII-(E)-153 (n. 8); "Remarks by Deputy Task Force Commander for Aviation," in CJTF-1, "Report," II-(B)-3 (n. 5); Moll, "Operation Crossroads," 66 (n. 19).

35. "Radiological Safety Plan Test Able," App. II to "Safety Plan," Annex E to CJTF-1, Operation Plan no. 1–46, E-II-2, -3 (n. 15); "Air Operations," VII-(E)-173, -174 (n. 8); "Special Reports: Safety," VII-(C)-11 (n. 22); "Chronology of ABLE DAY Operation," VI-(B)-8, -9 (n. 5); Bradley, *No Place to Hide*, 56–57 (n. 1).

36. Bradley, *No Place to Hide*, 58–59 (n. 1).

37. "Chronology of ABLE DAY Operation," VI-(B)-8 to -10, -14, -15 (n. 5); "Special Reports: Safety," VII-(C)-11, -12 (n. 22); "Ship Operations," VII-(F)-4, -31 (n. 25); Shurcliff, *Bombs at Bikini*, 128–129 (n. 9); Moll, "Operation Crossroads," 66 (n. 19).

38. "Reboarding and Inspection Plan," Annex X to CJTF-1, Operation Plan no. 1–46, X-4 (n. 15); "Chronology of ABLE DAY Operation," VI-(B)-11, -15 (n. 5); "Special Reports: Safety," VII-(C)-11, -12 (n. 22); Bradley, *No Place to Hide*, 60 (n. 1).

39. Stafford Warren to Viola L. Warren, 2 July 1946 (Stafford Warren Papers).

40. RSS to Tech. Dir., "Nuclear Radiation Effects in Tests Able and Baker, Preliminary Report of," 25 Sept. 1946, 7–8 (LASL Records); DSM, "Radiological Decontamination of Target and Non-Target Vessels," by Cmdr. J. J. Fee, 3 vols., Technical Report XRD-185, -186, -187, 1946, 1:3.

41. Capt. W. S. Maxwell to DSM, "SAKAWA—Condition of and Sinking," 2 July 1946 (LASL Records); RSS Briefing Sheet, 3 July 1946 (Stafford Warren Papers); "Chronology of ABLE DAY Operation," VI-(B)-15 to -18 (n. 5); "Special Reports: Safety," VII-(C)-14, -15 (n. 22); "Conduct of the Tests from the Operational Aspect," in CJTF-1, "Report," V-(C)-6 (n. 5).

42. Dessauer, "Instructions" (n. 10); Dessauer, "Personnel Film Badges [Able]," 22 July 1946; Dessauer, "Photographic Dosimetry," excerpts from a lecture, Aug. 1947, 3–4 (both in REECo microfilm records, roll 1); "Special Reports: Safety," VII-(C)-13 (n. 22). Cf. L. Berkhouse, S. E. Davis, F. R. Gladeck, J. H. Hallowell, C. B. Jones, E. J. Martin, F. W. McMullan, and M. J. Osborne, *Operation Crossroads, 1946*, report no. DNA 6032F (Washington, 1 May 1984), 216–217.

43. Dessauer, "Personnel Film Badges [Able]" (n. 42).

44. "Notes on JTF-1 Conference on Test Baker on Board U.S.S. *Mt. McKinley*, 10 July, 1946," 12 July 1946 (LASL Records).

45. "Resume of Able Day Operations," 6 July (n. 26). Cf. "Air Operations," VII-(E)-159 (n. 8).

46. DSM to Parsons, "Situation Report #20 Able," 7 July 1946 (LASL Records); DSM to RSS, 15 July 1946 (in "Crossroads Dispatches" folder, Crossroads box 47, DASIAC Library, Kaman Tempo, Santa Barbara, Cal.; hereafter cited as Crossroads Dispatches); "Roster of Job Assignments for the Radiological Safety Section for Test BAKER," 21 July 1946 (Stafford Warren Papers); "Chronology of ABLE DAY Operation," VI-(B)-18, -19 (n. 5); "Conduct of the Tests," V-(C)-4, -5, -7 (n. 41.)

47. "General Considerations—Radiological Safety—Test Baker," and "Radiological Safety Plan Test Baker," Change no. 7, 15 July 1946, App. IX and App. X, respectively, to "Safety Plan" (n. 35).

48. RSS to CJTF-1, 13 July 1946 (Crossroads Dispatches).

49. "Conduct of the Tests," V-(C)-7 (n. 41); "Chronology of WILLIAM DAY Operation," in CJTF-1, "Report," VI-(C)-5ff. (n. 5); "Aerological Report," 23 (n. 2); Moll, "Operation Crossroads," 67 (n. 19).

50. "Air Operations," VII-(E)-191, -192 (n. 8); "Aerological Report," 26 (n. 2); Cumberledge, "Aerological Aspects," 144–145 (n. 2).

51. Kirkpatrick diary, entry for 25 July 1946 (n. 20).

52. Percy W. Bridgman (Physics Dept., Harvard University) to Hans A. Bethe (Physics Dept., Cornell University), n.d., as quoted in Norris E. Bradbury to Maj. Gen. Leslie R. Groves, 131545Z, 13 March 1946 (LASL Records); Bethe to Bridgman, 13 Feb. 1946; Col. H. C. Gee to Bradbury, 121642Z, 12 March 1946; "General Considerations—Radiological Safety—Test Baker," E-IX-1 (n. 47); Kirkpatrick diary, entry for 25 July 1946 (n. 20); Bradley, *No Place to Hide*, 49–50, 86–87 (n. 1).

53. "Chronology of BAKER DAY Operation," in JTF-1, "Report," VI-(D)-6 (n. 5); President's Evaluation Commission, "Preliminary Report on the Second Bikini Atom Bomb Test," 29 July 1946; "Announced United States Nuclear Tests, July 1945 through December 1980," Office of Public Affairs, Nevada Operations Office, U.S. Department of Energy, report no. NVO-209 (Rev. 1), (Las Vegas, Jan. 1981), 2.

54. Kirkpatrick diary, entry for 25 July 1946 (n. 20).

55. Bradley, *No Place to Hide*, 92–93 (n. 1).

56. Shurcliff, "Technical Report," 22.3 (n. 9).

57. JCS Evaluation Board, "Preliminary Report [Baker]," Sect. II: "Observations on Test 'B' " (n. 34); Cumberledge, "Aerological Aspects," 147 (n. 2); Shurcliff, "Technical Report," 22.3 (n. 9); Samuel Glasstone, ed., *The Effects of Atomic Weapons* (Washington: Government Printing Office, 1950), 37–44; Armed Forces Special Weapons Project, *Radiological Defense*, vol. 2: *The Principles of Military Defense against Atomic Weapons* (Washington, Nov. 1951), 7, 9; Moll, "Operation Crossroads," 67 (n. 19).

58. Kirkpatrick diary, entry for 25 July 1946 (n. 20).

59. "Radiological Safety Plan Test Baker," E-X-4, -5 (n. 47); "Special Reports: Safety, " VII-(C)-19, -20 (n. 22); "Air Operations," VII-(E)-218, -219 (n. 8); "Conduct of the Tests," V-(C)-8 (n. 41); "Chronology of BAKER DAY Operation," VI-(D)-9 (n. 53).

60. Bradley, *No Place to Hide*, 96–97 (n. 1).

61. CJTF-1 to War Dept., "Situation Summary M Plus 2 Hours," 25 July 1946 (LASL Records); J. G. Ripczinske, "BAKER DAY Radiological Safety Notes," 25 July 1946, 2–3 (Stafford Warren Papers); Karl Z. Morgan to Stafford Warren, "Condition of Drone Boats," n.d. (Stafford Warren Papers); "Chronology of BAKER DAY Operation," VI-(D)-10 -18, -19 (n. 53); "Special Reports: Safety," VII-(C)-21, -22 (n. 22); "Conduct of the Tests," V-(C)-8 (n. 41).

62. Gerhard Dessauer, "Test Baker: Report on Personnel Monitoring by Means of Film Badges," 13 Aug. 1946; "Doses Received on Baker Day—25 July 46" (both in REECo microfilm records, roll 1). Cf. Berkhouse et al., *Operation Crossroads*, 222: "Table 23. Film badge summary (in roentgens) of radiological patrol boat crews, CROSSROADS" (n. 42).

63. "General Considerations—Radiological Safety—Test Baker" (n. 47); Shurcliff, *Bombs at Bikini*, 159–160 (n. 9); Walmer E. Strope, "Investigation of Gamma Radiation Hazards Incident to an Underwater Atomic Explosion," BuShips Preliminary Design Section, March 1948; Glasstone, ed., *Effects of Atomic Weapons*, 276–283 (n. 57); *Radiological Defense*, vol. 2, 11, 54–57 (n. 57).

64. CJTF-1 to JCS, 251211Z, 25 July 1946, as quoted in "Chronology of BAKER DAY Operation," VI-(D)-18 (n. 53).

65. CJTF-1 to JCS, 25 July (n. 64); DSM to Tech. Dir., "Situation Report No. 29," 25 July 1946, as quoted in "Chronology of BAKER DAY Operation," VI-(D)-18, -19 (n. 53); "Conduct of the Tests," V-(C)-8, -9 (n. 41); DSM, "Technical Inspection Report," 28–48 (n. 29).

66. Stafford Warren to CJTF-1, "Summary Report of Conditions of Target Ships as of 2000 [8:00 p.m.] 9 August 1946 Based on DSM Plot in Radiological Safety Control," 9 Aug. 1946 (Stafford Warren Papers).

67. Stafford Warren to Viola Warren, 26 July 1946 (Stafford Warren Papers).

68. CJTF-1 to JCS, 260918Z, 26 July 1946 (LASL Records).

69. CJTF-1 to JTF-1, 270711Z. 27 July 1946 (LASL Records).

70. CJTF-1 to JCS, 280838Z, 28 July 1946 (LASL Reccords).

71. CJTF-1 to JCS, 300913Z, 30 July 1946, as quoted in "Chronology of BAKER DAY Operation," VI-(D)-44 (n. 53); "Post Test Activities," in CJTF-1, "Report," V-(D)-1, -2 (n. 5).

72. Shurcliff, "Technical Report," 21.4 (n. 9).

73. Ibid., 25.3.

74. Stafford Warren to CJTF-1, "Review of Radiological Safety Situation," 3 Aug. 1946, 1 (Stafford Warren Papers).

75. RSS, Memorandum on the radiological status of the target ships, 5 Aug. 1946 (Stafford Warren Papers).

76. DSM to CJTF-1, 291025Z, 29 July 1946, as quoted in "Chronology of BAKER DAY Operation," VI-(D)-40 (n. 53); DSM to All Target Vessels, "Decontamination Procedures on Target Vessels," 31 July 1946, in DSM, "Radio-

logical Decontamination," 3:4–7 (n. 40); RSS, "Radiation Intensity and Permitted Working Time at Various Rates," table in Warren to CJTF-1, 3 Aug. (n. 74); DSM to Commander, Task Unit 1.2.7, et al., "Preliminary Decontamination of Target Vessels by Ships of TU 1.2.7," 4 Aug. 1946, in DSM, "Radiological Decontamination," 3:8–13; "Special Reports: Safety," VII-(C)-24 (n. 22); "Post Test Activities," V-(D)-1 (n. 5), Bradley, *No Place to Hide*, 102–103 (n. 1); Berkhouse et al., *Operation Crossroads*, App. F. "Radiation Readings Aboard Target Vessels" (n. 42).

77. RSS, "Tabulation of Decay Time from Various Intensities down to 0.1r/day," table in Warren to CJTF-1, 3 Aug. (n. 74); RSS, "Nuclear Radiation Effects," App. 7, "Radiological Situation on Target Ships" (n. 40); DSM, "Radiological Decontamination," pt. 1, "Decontamination of Target Vessels," and App. 1, "Reports of Target Vessel Decontamination" (n. 40). More generally, see Glasstone, ed., *Effects of Atomic Weapons*, chap. 10, "Decontamination" (n. 57).

78. CJTF-1 to JCS, 270851Z, 27 July 1946 (LASL Records).

79. CJTF-1 to JCS, 280838Z, 28 July 1946 (LASL Records).

80. Stafford Warren to Viola Warren, 30 July 1946 (Stafford Warren Papers).

81. Stafford Warren, "Review of Radiological Safety Situation," 3 (n. 74).

82. Stafford Warren, "Radiological Safety Regulations," n.d. Cf. Commander Task Group [CTG] 1.2 for distribution, "Processing of Working Parties Proceeding to and from Work on Contaminated Target Vessels—Instructions for," 13 Aug. 1946; Lt. J. J. Castelvecchi (Repair Officer), approved Lt. Thomas A. Shallow (RSS Monitor), "U.S.S. AJAX (AR-6), Decontamination Procedure," n.d. (all in Stafford Warren Papers).

83. RSS Medico-Legal Board, minutes of meetings on 2 and 3 Aug. 1946, 3 Aug. 1946 (Stafford Warren Papers); Karl Z. Morgan, "Final Report of the Alpha Beta Gamma Survey Section," 6 Aug. 1946 (Stafford Warren Papers); CTG 1.2 to Task Group 1.2, 091244Z, 9 Aug. 1946 (Crossroads Dispatches); Stafford L. Warren, "The Role of Radiology in the Development of the Atomic Bomb," in Kenneth D. A. Allen, ed., *Radiology in World War II*, Medical Department in World War II: Clinical Series (Washington: Office of the Surgeon General, U.S. Army, 1966), 911, 913.

84. Herbert Scoville, Jr., to JTF-1 Asst. Chief of Staff for Logistics, "Radiological Hazard from Contamination of Ships' Evaporators and Condensers," 27 April 1946; RSS, "Radiological Monitoring of Evaporators," July 1946; RSS, Committee on Evaporators, "Recommendations," 8 Aug. 1946; Ens. William A. Loeb to Stafford Warren, "Monitor Instruction, Evaporators and Ventilators," 12 Aug. 1946; Ens. Howard R. Walton, "Ventilation Problems and Clearance," n.d.; "Collecting Samples from Ventilators," n.d. (all in Stafford Warren Papers); Bradley, *No Place to Hide*, 103–105 (n. 1). Cf. DSM, "Radiological Decontamination," 18–20 (n. 40).

85. Lyon to Chief, Bureau of Medicine and Surgery [BuMed], "Marine Biology, Consultant in, Appointment of," 27 Aug. 1946 (Stafford Warren Papers); Neal O. Hines, *Proving Ground: An Acount of the Radiobiological Studies in the Pacific, 1946–1961* (Seattle: University of Washington Press, 1962), 45, 48.

86. Destroyer DD-692 *Allen M. Sumner* to CJTF-1, 281252Z, 28 July 1946, as quoted in "Chronology of BAKER DAY Operation," VI-(D)-36 (n. 53).

87. CJTF-1 to *Sumner*, 281246Z, 28 July 1946, as quoted in "Chronology of BAKER DAY Operation," VI-(D)-36 (n. 53). Cf. CJTF-1 to All Ships Present at Bikini, 311521Z, 31 July 1946; *Sumner* to DSM, 030445Z, 3 Aug. 1946 (both in Crossroads Dispatches).

88. DSM, "Radiological Decontamination," 1:20 (n. 40).

89. Morgan, "Final Report," 1–2 (n. 83).

90. Stafford Warren to Viola Warren, 2 Aug. 1946 (Stafford Warren Papers).

91. Dessauer, "Test Baker" (n. 62); "Special Reports: Safety," VII-(C)-24 (n. 22); "Post Test Activities," V-(D)-4 (n. 71); Dessauer, "Photographic Dosimetry," 4 (n. 42). Cf. Berkhouse et al., *Operation Crossroads*, 219: "Table 20. CROSSROADS badging after shot BAKER" (n. 42).

92. Stafford Warren, "Review of Radiological Safety Situation," 2 (n. 74). Cf. Lt. Robert J. Rieckhoff and Lt. (jg) Delbert W. Jones (RSS Monitors for *Prinz Eugen*) to Lt. Cmdr. William A. Wulfman, "Radiological Conditions Aboard the USS Prinz Eugen," 10 Aug. 1946; RSS Medico-Legal Board, minutes of meeting about radiation exposures of working parties on USS *Salt Lake City*, 10 Aug. 1946 (Stafford Warren Papers).

93. Stafford Warren to CJTF-1, "Occupancy of Target Vessels as Influenced by Intensity of Radiation of Various Types on Target Vessels," 7 Aug. 1946, 1 (Stafford Warren Papers).

94. Medico-Legal Board minutes, 3 Aug., items 1–2 (n. 83).

95. Warren to CJTF-1, 7 Aug., 2 (n. 93). Cf. William G. Myers to Stafford Warren, "Observations and Suggestions Resulting from Participation in Test Able and Test Baker," 27 Aug. 1946.

96. Warren to CJTF-1, 7 Aug., 3 (n. 93).

97. Stafford Warren, "Review of Radiological Safety Situation," 3 (n. 74).

98. Lt. Ralph D. Ross to All Personnel of Radiological Safety Section, "Personnel Remaining at Bikini Atoll during 'Stabilization' Period, Crossroads Operation," and "Personnel Returning to the USA via USS HENRICO (APA-45)," 13 Aug. 1946 (Stafford Warren Papers).

99. Warren to CJTF-1, 7 Aug., 3–4 (n. 93); Myers to Stafford Warren, 27 Aug. (n. 95); Donald L. Collins, ed., "Operations Crossroads Reports to Rad Safe Instrument Division," n.d. (both in Stafford Warren Papers).

100. Warren, "Review of Radiological Safety Situation," 4 (n. 74)

101. RSS Medico-Legal Board, minutes of meeting, 4 Aug. 1946 (Stafford Warren Papers).

102. DSM, "Radiological Contamination," 1:13 (n. 40).

103. DepComTech to DSM and RSS, "DepComTech Highlight Summary no. 24," 060554Z, 6 Aug. 1946 (LASL Records).

104. Warren to CJTF-1, 7 Aug., 1 (n. 193).

105. Stafford Warren to Viola Warren, 11 Aug. 1946 (Stafford Warren Papers). Cf. Bradley, *No Place to Hide*, 104, 109–114 (n. 1).

106. Lyon to Parsons, "Ensign Coffin," 5 May 1947 (Stafford Warren Papers).

107. William Rubinson to Bradbury, "Some Remarks on the Decontamination of Radioactive Ships," 26 Aug. 1946 (LASL Records).

108. Rieckhoff and Jones to Wulfman, 10 Aug. (n. 92).

109. A. H. Graubart to CTG 1.2, "Radiological Decontamination Procedures, USS PRINZ EUGEN, August 4 to August 11, 1946," 13 Aug. 1946, 4. Cf. R. J. Connell (Commanding Officer, USS *New York*) to DSM, "Report of Decontamination Progress, U.S.S. NEW YORK," 15 Aug. 1946, 4 (both in Crossroads Dispatches).

110. RSS, "Nuclear Radiation Effects," 10 (n. 40).

111. Radiological Safety Advisor to Capt. S. P. Bednarczyk, 130820Z, 13 Aug. 1946 (Stafford Warren Papers). Cf. M. G. Bowman (LASL Radio-Chemistry Group), "Sampling Deck of New York for Alpha Contamination (with Notes on the Alpha Contamination Problem)," 7 Aug. 1946 (Stafford Warren Papers); Rubinson to Bradbury, 26 Aug. (n. 107); DSM, "Radiological Decontamination," 1:13 (n. 40); Berkhouse et al., *Operation Crossroads*, 116–120 (n. 42).

112. Warren to CJTF-1, 7 Aug., 2 (n. 93). Cf. RSS Medico-Legal Board, minutes of meetings on 12 and 13 Aug. 1946; Stafford Warren to Groves, "Clearance of 'Discussion on Product [Plutonium] Contamination after Underwater Bomb Detonation,' during Part of Informal Talks on Safety at Bikini," 9 Oct. 1946 (both in Stafford Warren Papers).

113. Warren to CJTF-1, 7 Aug., 4–5 (n. 93). Cf. CJTF-1 to Commanding General, Manhattan District, 090826Z, 13 Aug. 1946; Derry to Bradbury, 141622Z, 14 Aug. 1946 (both in LASL Records).

114. Warren to Warren, 11 Aug. (n. 105). Cf. Stafford Warren to Rear Adm. F. G. Fahrion, memo on plutonium contamination of fish in Bikini lagoon, 13 Aug. 1946 (Stafford Warren Papers); DSM, "Radiological Decontamination," 1:13–14 (n. 40).

115. Stafford Warren to CTG 1.2, memo on alpha contamination of target ships, 13 Aug. 1946 (Stafford Warren Papers).

116. CJTF-1 to CincPac/CincPoa, 100648Z, 10 Aug. 1946 (Crossroads Dispatches). Cf. "Special Reports: Safety," VII-(C)-23 (n. 22).

117. Stafford Warren to Viola Warren, 15 Aug. 1946 (Stafford Warren Papers).

118. "Post Test Activities," V-(D)-2, -3, -5 (n. 71); Shurcliff, *Bombs at Bikini*, 205–206 (n. 9); Bradley, *No Place to Hide*, 115 (n. 1); Moll, "Operation Crossroads," 68 (n. 19).

119. The President to the Secretary of the Navy, 7 Sept. 1946 (in President's Secretary File, Papers of Harry S. Truman, Truman Library, Independence, Mo.; cited hereafter as Truman Papers, Sec. File). Cf. Parsons to Tech. Dir. et al., "Preparations for Test Charlie," 13 Sept. 1946 (LASL Records).

120. Lyon to Vice Adm. Ross T. McIntire (Chief, BuMed), letter report on Crossroads, 12 Aug. 1946, 2–4; RSS, "Radiological Monitoring of Ships Returning from Operation 'Crossroads,' " n.d.; CJTF-1 to Commanding Officer of Ships Listed, "Monitoring of Ships to Check Radiological Contamination," 19 Aug. 1946 (all in Stafford Warren Papers); "Post Test Activities," V-(D)-4 (n. 71); DSM, "Radiological Decontamination," 1:20–21 (n. 40). Cf. Berkhouse et

al., *Operation Crossroads*, chap. 5, "Post-Baker Operations: Bikini, Kwajalein, and the United States," and App. A, "Activities of Participating Navy Vessels during Operation Crossroads" (n. 42).

121. ComServPac to Commander, Western Sea Frontier [CWSF], et al., 240111Z, 24 Aug. 1946 (Crossroads Dispatches); Radiological Safety Advisor (Warren) to AdCom, CJTF-1, 252345Z, 25 Aug. 1946; Safety Advisor (Lyon) to CJTF-1, 032141Z, 3 Sept. 1946; Safety Advisor to RadSafe Rep and CJTF-1, 032142Z, 3 Sept. 1946; Safety Advisor to Tech. Dir., 032143Z, 3 Sept. 1946 (LASL Records); CJTF-1 to Commanding Officers of Ships Listed, "Ships, Radiological Safety of," 9 Sept. 1946, as reproduced in DSM, "Radiological Decontamination," 1:125–145 (n. 40); DSM, "Radiological Decontamination," 1:21–25 (n. 40).

122. DSM, "Radiological Decontamination," 1:24 (n. 40).

123. Chief of Naval Operations [CNO] to All Bureaus, Boards and Officers of the Navy Department, "Establishment of a Radiological Safety Program for the Navy," 27 Aug. 1946, as reproduced in DSM, "Radiological Decontamination," 3:119–121 (n. 40).

124. Chief, BuShips, to CWSF et al., "Radiological Decontamination Program, Development of," 4 Nov. 1946; Chief, BuShips, to Commander, San Francisco Naval Shipyard [CSFNS], "Laboratory for Radiological Studies; Supplement to Present Laboratory Facilities for," 18 Nov. 1946, as reproduced, respectively, in DSM, "Radiological Decontamination," 2:114–116, 3:117–118 (n. 40); ibid., App. 3, "Non-Target Vessel Experimental Decontamination Work" (n. 40); "Post Test Activities," V-(D)-5 (n. 71).

125. Philip Lemler to CSFNS, "Summary of Rear Admiral Solberg's Visit to San Francisco Naval Shipyard on Tuesday, 17 September 1946, Regarding Experimental Work on the Bikini Radioactive Ships" and "Summary of . . . Visit . . . on Wednesday, 18 September . . . ," 19 Sept. 1946; Solberg, "Report of Activities in San Francisco Area from Tuesday, 17 September to Friday, 20 September," 20 Sept. 1946; BuShips and BuMed to CincPac et al., "Radiological Clearance of Non Target Vessels and Procedures for Decontamination," 24 Sept. 1946, all as reproduced, respectively, in DSM, "Radiological Decontamination," 2:27–28, 2:29–32; 2:35–38; and 3:16–19 (n. 40).

126. BuShips and BuMed, "Radiological Clearance," 18–19 (n. 125). Cf. BuShips and BuMed to CWSF, 141550Z, 14 Oct. 1946, as reproduced in DSM, "Radiological Decontamination," 3:22–23 (n. 40).

127. "DepComTech Highlight Summary no. 24," 6 Aug. (n. 103); Lyon to McIntire, 12 Aug., 5–7 (n. 120); Stafford Warren to Viola Warren, 20 Aug. 1946 (Stafford Warren Papers); Safety Advisor (Lyon) to CJTF-1, 202345Z, 21 Aug. 1946 (Crossroads Dispatches); Blandy to Army Chief of Staff, "Radiological Safety—Training of Personnel," 27 Aug. 1946 (LASL Records); JTF-1, Radiological Safety School, "Transcript of Conference," 9 Oct. 1946 (Stafford Warren Papers); Lyon to CJTF-1, "Radiological Safety Plan of Organization within the Navy," 11 Oct. 1946 (Stafford Warren Papers); Cmdr. J. J. Fee for files, "Radiological Decontamination; Conference Concerning," 14 Oct. 1946, pars. 7–12, as reproduced in DSM, "Radiological Decontamination," 3:77–83; Shurcliff, "Technical Report," 31.5 (n. 9).

128. Cmdr. Draper L. Kauffman, "Report of the Radiological Safety School, Joint Task Force One, Navy Department, Washington, 9 September 1946 through 9 October 1946," 14 Oct. 1946; Lt. Cmdr. Arthur B. Cutts, "Report of the Radiological Safety School, Bureau of Medicine and Surgery, Navy Department, Washington, 25 November 1946 through 9 December 1946," 20 Dec. 1946 (both in Stafford Warren Papers).

129. Fee to Capt. W. E. Walsh, 102030Z, 10 Oct. 1946, as reproduced in DSM, "Radiological Decontamination," 3:74 (n. 40); CWSF to BuShips, 112327Z, 11 Oct. 1946, ibid., 3:76; Lyon to Walsh, "Tolerance Figures," 18 Oct. 1946, ibid., 3:84; BuMed to Walsh, 252116Z, 25 Oct. 1946, ibid., 3:94; CWSF to Chief, BuMed, 060145Z, 6 Nov. 1946, ibid., 3:97–98; CWSF to BuMed, 151807Z, 15 Nov. 1946, ibid., 3:90; BuMed to CWSF, 222127Z, 22 Nov. 1946, ibid., 3:29.

130. As quoted in Fee for files, 14 Oct., pars. 14(b), 16 (n. 127).

131. BuShips and BuMed to Distrib. List, "Radiological Clearance and Decontamination Procedures for Crossroads Non-Target Vessels," 22 Nov. 1946, pars. 5–7, as reproduced in DSM, "Radiological Decontamination," 3:30–51, at 32–33 (n. 40).

132. BuShips and BuMed, "Radiological Clearance," 22 Nov., par. 2 (n. 131).

133. Bradbury to Gee, 232050Z, 23 Aug. 1946 (LASL Records); R. R. Coveyou to Stafford Warren, "Gamma-Alpha Equivalences for Flat Surfaces," 26 Aug. 1946 (Stafford Warren Papers); Radiological Safety Advisor to Capt. Bednarczyk, 260200Z, 26 Aug. 1946 (Stafford Warren Papers); Bradbury to Gee, letter on decontamination problems, 24 Sept. 1946 (LASL Records); Fee for files, 14 Oct., par. 6 (n. 127); Fee to Walsh, 012219Z, 1 Nov. 1946, as reproduced in DSM, "Radiological Decontamination," 3:88 (n. 40); Rear Adm. Clifford A. Swanson (Chief, BuMed) to the Surgeon General, U.S. Public Health Service, "Request for the Services of the United States Public Health Service Personnel for Work in Connection with Operation CROSSROADS," 11 Dec. 1946 (Stafford Warren Papers); Joint Crossroads Committee, "Radiological Safety," n.d., published as *Radiological Defense* (n. 57), vol. 1 (Washington, 1948), chap. 9, "Principles of Instrumentation and Photographic Dosimetry." Cf. F. R. Shonka and H. L. Wyckoff, "Measurement of Nuclear Radiations," in Glasstone, ed., *Effects of Atomic Weapons*, 291–311 (n. 57).

134. Comment by Capt. W. H. Maxwell, in "Report of Conference, San Francisco Naval Shipyard," 1 Oct. 1946, as reproduced in DSM, "Radiological Decontamination," 3:61–73, at 65.

135. CWSF to Commandant Eleventh Naval District et al., "Radiological Monitoring Organization," 18 Nov. 1946, as reproduced in DSM, "Radiological Decontamination," 3:26–28 (n. 40); BuShips and BuMed, "Radiological Clearance," w/encls. A, "General Radiological Safety Precautions, Crossroads Vessels," and B, "Radiological Decontamination Procedures" (n. 131).

136. Fee, "Conference on Radiological Safety; Report of," 10 Dec. 1946, par. 1, as reproduced in DSM, "Radiological Decontamination," 3:102–115 (n. 40).

137. Fee, "Conference," Pars. 11–12 (n. 136); BuShips to CWSF et al., 032133Z, 4 Dec. 1946, as reproduced in DSM, "Radiological Decontamination," 3:53 (n. 40); BuShips and BuMed to Distrib. List, "Radiological Clearance and

Decontamination Procedures for CROSSROADS Non-Target Vessels," 17 Dec. 1946, ibid., 3:54–58.

138. As quoted in Fee, "Conference," par. 11 (n. 136).

139. As quoted in ibid., par. 18.

140. CJTF-1 to CNO, 11 Aug. 1946 (Crossroads Dispatches); CTG 1.2 to USS *Conyngham* et al., 120706Z, 12 Aug. 1946 (Crossroads Dispatches); Cmdr. C. L. Gaasterland to DSM, "Comments on Radiation Measurements on Target Submarines," 14 Aug. 1946 (Stafford Warren Papers); Commander, Naval Task Groups [CNTG], to AdComd CJTF-1, 221156Z, 22 Aug. 1946 (LASL Records); ComServPac to CWSF et al., 240111Z, 24 Aug. 1946 (Crossroads Dispatches); Fahrion to CJTF-1, "Weekly Report for Week Ending 24 August, 1946," 25 Aug. 1946 (Crossroads Dispatches); "Post Test Activities," V-(D)-2 (n. 71).

141. "Summation by Commander Joint Task Force One," in CJTF-1, "Report," II-(A)-3 (n. 5).

142. CincPac to AlPac 238, 032333Z, 3 Sept. 1946 (LASL Records); CNTG to All USNav Ships at Kwaj, 051648Z, 5 Sept. 1946 (LASL Records); "Post Test Activities," V-(D)-6, -7 (n. 71).

143. Fahrion to CJTF-1, "Weekly Report for Week Ending 7 September, 1946," 8 Sept. 1946, 3–4 (Crossroads Dispatches).

144. Capt. Frank I. Winant, Jr., "Command Aspects of Radiological Defense," in *Radiological Defense* (n. 57), vol. 3: *A Series of Indoctrination Lectures on Atomic Explosion, with Medical Aspects* (Washington, n.d.), 106. Cf. Winant to Chief, BuMed, "Radiological Safety," 11 Nov. 1946; Lyon, "Comments on Letter of Officer-in-Charge of Ammunition Disposal Unit of 11 November 1946," 29 Nov. 1946 (both in Crossroads Dispatches).

145. Fahrion to CJTF-1, "Weekly Report for Week Ending 19 October 1946," 20 Oct. 1946 (Crossroads Dispatches); "Dosimetry Matrix Report, 1946 Pacific Records," REECo, compiled 22 July 1982; Berkhouse et al., *Operation Crossroads*, 218, 220 (n. 42).

146. Scoville to Stafford Warren, "Residual Radioactivity on CROSSROADS Target Vessels," 3 Jan. 1947. Cf. Radiological Safety Advisor to Capt. Bednarczyk, 040158Z, and 041210Z, 4 Oct. 1946 (Stafford Warren Papers).

147. Stafford Warren to Parsons, "Hazards from Residual Radioactivity on the Crossroads Target Vessels," 6 Jan 1946 (Stafford Warren Papers).

148. CJTF-1 to Commanding Officer, USS *Severn* (AO-61), et al., "Ships, Radiological Safety of," 26 Sept. 1946 (Crossroads Dispatches); Fahrion, "Weekly Report," 20 Oct., 1 (n. 145); "Post Test Activities," V-(D)-6 (n. 71); Chief, BuShips, to Commander, Pearl Harbor Naval Shipyard, et al., "Radiological Examination of CROSSROADS Target Ships," 14 Jan. 1947, as reproduced in DSM, "Radiological Decontamination," 3:122–126 (n. 40); Chief, BuMed, to Distrib. List, "Safety Regulations for Work in Target Vessels formerly JTF-1," 31 Jan. 1947 (Crossroads Dispatches).

149. "Remarks by the Deputy Task Force Commander for Technical Direction," in CJTF-1, "Report," II-(C)-3 (n. 5).

150. RSS to Tech. Dir., "Densitometric Film Badges (Gamma Ray-Dose) Preliminary Survey," 10 July 1946 (Stafford Warren Papers); RSS, "Nuclear Radia-

tion Effects," 10–11, 16 (n. 40); R. H. Draeger and Shields Warren, "Medicine at the Crossroads," *U.S. Naval Medical Bulletin* 47 (1947):219–225.

151. JCS Evaluation Board, "Preliminary Report [Baker]" (n. 34).

152. Shurcliff, "Technical Report," 27.6 (n. 9).

153. JCS Evaluation Board, "Preliminary Report [Baker]" (n. 34). Cf. President's Evaluation Commission, "Preliminary Report [Baker]" (n. 53).

154. RSS, "Nuclear Radiation Effects," 10–14, 16–17 (n. 40); DSM, "Technical Inspection Report," sec. 4, "Radiological Effects—Summary and General Conclusions" (n. 29); John C. Tullis, "The Gross Autopsy Findings and a Statistical Study of the Mortality in the Animals Exposed at Bikini," DSM Naval Medical Research Section, "Appendix no. 3 to the Final Report," 1 March 1947; Tullis and Shields Warren, "Gross Autopsy Observation in the Animals Exposed at Bikini: A Preliminary Report," *Journal of the American Medical Association* 134 (1947):1155–1158; Shurcliff, *Bombs at Bikini*, 140–144, 166–168 (n. 9).

155. As quoted in Tris Coffin, "In the Light of Bikini," *Nation* 163 (5 Oct. 1946):371.

156. Draeger and Warren, "Medicine at the Crossroads" (n. 150); Coffin, "In the Light of Bikini," 370–371 (n. 155); Hines, *Proving Ground*, chap. 2 (n. 85).

157. Stafford Warren to Viola Warren, 13 Aug. 1946 (Stafford Warren Papers).

158. Lyon to McIntire, 12 Aug., 1 (n. 120).

159. Myers to Stafford Warren, 27 Aug., 5 (n. 95).

160. Stafford Warren to Myers, 31 Dec. 1946 (Stafford Warren Papers). The apparent tardiness of this response was actually the result of Warren's failure to receive the original report; see Myers to Warren, 9 Dec. 1946 (Stafford Warren Papers).

161. Robert R. Newell, "Report of the Medic[o]-Legal Board," 19 Aug. 1946 (Stafford Warren Papers).

162. "Summation by Commander," II-(A)-2 to -4 (n. 141; quoted at -2).

163. "Operation CROSSROADS," DNA Public Affairs Office Fact Sheet, 5 April 1984, 2. Cf. Berkhouse et al., *Operation Crossroads*, 2–3 (n. 42).

164. Harvey Wasserman and Norman Solomon, with Robert Alvarez and Eleanor Walters, *Killing Our Own: The Disaster of America's Experience with Atomic Radiation* (New York: Dell, 1982), 37–46; Michael Uhl and Tod Ensign, *GI Guinea Pigs: How the Pentagon Exposed Our Troops to Dangers More Deadly Than War: Agent Orange and Atomic Radiation* (Chicago: Playboy Press, 1980), chap. 2; Arjun Makhijani and David Albright, "Irradiation of Personnel during Operation Crossroads: An Evaluation Based on Official Documents," International Radiation Research and Training Institute, Washington, May 1983; Philip H. Melanson, "The Human Guinea Pigs at Bikini," *Nation*, 9–16 July 1983, 1, 48–50.

NOTE ON SOURCES

In writing this book I relied chiefly on the documentary residue of the Manhattan Project and related programs, which I obtained from a number of repositories. Of major importance for my subject were collections in the Records Center, Argonne National Laboratory, Argonne, Illinois; Mail and Records Center, Los Alamos National Laboratory, Los Alamos, New Mexico; President's Secretary File, Papers of Harry S. Truman, Truman Library, Independence, Missouri; E. O. Lawrence Collection, Bancroft Library, University of California, Berkeley, California; Stafford Warren Papers, University Research Library, University of California, Los Angeles, California; and Crossroads Dispatches File, DASIAC Library, Kaman Tempo, Santa Barbara, California.

Other record centers and archives have also afforded me a wealth of documents through the efforts of the U.S. Department of Energy Coordination and Information Center, which opened in 1981. Reynolds Electrical & Engineering Co., Inc., Las Vegas, Nevada, operates CIC, which has collected material relevant to radiation safety in nuclear weapons testing from a number of federal record centers and other archives. Copies of all these documents are unclassified or declassified and are publicly available at the CIC in Las Vegas. Such material, for the most part, does not lend itself readily to bibliographical listing. In the bibliography I include all published works cited in the footnotes but only a few of the more important unpublished reports and other items.

Although resting chiefly on archival sources, this book also draws on the oral testimony of many participants and eyewitnesses. My interviews with surviving members of the Chicago Health Division and its offshoots at Argonne, Oak Ridge, and Hanford, the Los Alamos Health Group, and the Manhattan Engineer District Health Office, among others, have aided me considerably in understanding the documentary record. Many of those I interviewed have also commented on all or part of the manuscript. So have interested scientists, historians, other scholars, and laypersons. Their specific contributions are shown in the footnotes; their names appear in lists of interviews and reviewers following this note.

INTERVIEWS

Abrams, Richard. Pittsburgh, Pennsylvania. 9 May 1980.
Allen, J. Garrott. Stanford, California. 15 April 1980.
Auxier, John A. Oak Ridge, Tennessee. 20 July 1979.
Backman, George H. Richland, Washington. 2 October 1979.
Bailey, Jack. Oak Ridge, Tennessee. 19 July 1979.
Bernstein, George. Argonne, Illinois. 19 June 1979.
Bond, Victor P. Upton, New York. 15 August 1979.
Brady, William J. Mercury, Nevada. 25 May & 15 June 1978.
Brown, Bernard L. Mercury, Nevada. 24 February 1981.
Brues, Austin M. Argonne, Illinois. 19 June 1979.
Carlson, Oscar N. Ames, Iowa. 21 October 1980.
Chiotti, Premo. Ames, Iowa. 20 October 1980.
Cohn, Waldo E. Oak Ridge, Tennessee. 18 July 1979.
Cole, Kenneth S. San Diego, California. 1 November 1979.
Collins, Donald L. Glendale, California. 31 October 1979.
Conard, Robert A. Upton, New York. 15 August 1979.
Congdon, Charles C. Knoxville, Tennessee. 13 July 1979.
Cooper, Robert. Mercury, Nevada. 14 November 1978.
Cronkite, Eugene P. Upton, New York. 15 August 1979.
Dummer, Jerome E. Los Alamos, New Mexico. 10 June 1980.
Eberline, Howard C. Edmond, Oklahoma. 28 September 1978.
Eisenbud, Merril. Tuxedo, New York. 24 October 1979.
Evans, Robley D. Scottsdale, Arizona. 8 November 1978.
Farmakes, John. Argonne, Illinois. 20 June 1979.
Ferry, John L. East Chicago, Indiana. 13 May 1980.
Finkle, Raymond D. Los Angeles, California. 17 April 1980.
Friedell, Hymer L. Cleveland, Ohio. 12 June 1979.
Fuqua, Philip A. Richland, Washington. 3 October 1979.
Gamertsfelder, Carl C. Knoxville, Tennessee. 13 July 1979.
Hagen, Charles W., Jr. Bloomington, Indiana. 20 May 1980.
Harley, John H. New York, New York. 18 October 1979.
Healy, John W. Los Alamos, New Mexico. 5 June 1980.
Heid, Kenneth R. Richland, Washington. 1 October 1979.

Hempelmann, Louis H., Jr. Los Alamos, New Mexico. 3-4 June 1980.
Hurst, G. S. Oak Ridge, Tennessee. 20 July 1979.
Hutchens, Tyra T. Portland, Oregon. 26 September 1979.
Jacks, Gordon L. Las Vegas, Nevada. 19 February 1980.
Jacobson, Leon O. Chicago, Illinois. 14 June 1979.
Johnson, William S. Phoenix, Arizona. 9 November 1978.
Kathren, Ronald L. Richland, Washington. 4 October 1979.
Kennedy, William R. Los Alamos, New Mexico. 1 May 1979.
Kuper, J. B. Horner. Upton, New York. 16 August 1979.
Lawrence, James N. P. Los Alamos, New Mexico. 11 June 1980.
LeRoy, George V. Pentwater, Michigan. 13 June 1979.
Littlejohn, George J. Los Alamos, New Mexico. 12 June 1980.
Martens, John H. Argonne, Illinois. 20 June 1979.
Morgan, Karl Z. Atlanta, Georgia. 25 July 1979.
Newman, Robert W. Las Vegas, Nevada. 30 June 1982.
Nickson, James J. Memphis, Tennessee. 12 July 1979.
Nolan, James F. Los Angeles, California. 17 April 1980.
Norwood, W. Daggett. Richland, Washington. 3 October 1979.
Novak, John. North Riverside, Illinois. 14 June 1979.
Parker, Herbert M. Richland, Washington. 5 October 1979.
Peterson, David T. Ames, Iowa. 22 October 1980.
Pinson, Ernest A. Mt. Vernon, Indiana. 26 July 1979.
Prosser, C. Ladd. Urbana, Illinois. 31 August 1979.
Ray, Roger. Las Vegas, Nevada. 28 July 1982.
Reed, George W. Argonne, Illinois. 21 May 1979.
Rose, John E. Mountain Home, Arkansas. 11 July 1979.
Sacher, George A. Argonne, Illinois. 29 May 1979.
Schubert, Jack. Holland, Michigan. 13 June 1979.
Schwartz, Samuel. Minneapolis, Minnesota. 25 June 1979.
Scott, Arthur F. Portland, Oregon. 26 September 1979.
Shockley, Vernon E. Richland, Washington. 4 October 1979.
Simmons, Eric L. Chicago, Illinois. 14 May 1980.
Sinclair, Bruce A. Pittsburgh, Pennsylvania. 21 October 1978.
Spedding, Frank H. Ames, Iowa. 21 October 1980.
Spinrad, Bernard I. Corvallis, Oregon. 26 September 1979.
Stannard, J. Newell. San Diego, California. 19 September 1980.
Sterner, James H. Irvine, California. 16 April 1980.
Storm, Ellery. Los Alamos, New Mexico. 13 June 1980.
Taft, Robert W. Las Vegas, Nevada. 15 July 1982.
Tannenbaum, Albert. La Jolla, California. 2 November 1979.
Taylor, Lauriston S. Bethesda, Maryland. 23 October 1978.
Tompkins, Paul C. Nashua, New Hampshire. 21 August 1979.
Voelz, George L. Los Alamos, New Mexico. 6 June 1980.
Voigt, Adolf. Ames, Iowa. 23 October 1980.
Warren, Stafford L. Los Angeles, California. 30 October 1979.
Wattenberg, Lee W. Minneapolis, Minnesota. 26 June 1979.

Wilcox, Floyd W. Las Vegas, Nevada. 15, 17, and 23 January 1979.
Wilhelm, Harley A. Ames, Iowa. 22 October 1980.
Wollan, Ernest O. Edina, Minnesota. 4 September 1979.

REVIEWERS

NOTE: The names of persons identified in the preface or in the list of interviews are not here repeated.

Aragon, Jesse. Los Alamos National Laboratory. Los Alamos, New Mexico.
Badash, Lawrence. University of California. Santa Barbara, California.
Beardon, Alan. University of California. Berkeley, California.
Beck, W. L., Jr. Bechtel National, Inc. Oak Ridge, Tennessee.
Bernstein, Barton J. Stanford University. Stanford, California.
Bernstein, Geoffrey. Cambridge, Massachusetts.
Blanchett, Jeremy. Defense Nuclear Agency. Bethesda, Maryland.
Boren, Paul. Defense Nuclear Agency. Washington, D.C.
Bramlett, Walter R. Los Alamos National Laboratory. Los Alamos, New Mexico.
Campbell, John H. Nevada Operations Office. Las Vegas, Nevada.
Cape, J. D. Technical Information Center. Oak Ridge, Tennessee.
Carew, Paul H. Defense Nuclear Agency. Alexandria, Virginia.
Cook, C. Sharp. University of Texas. El Paso, Texas.
Dahl, Adrian H. Los Alamos National Laboratory. Los Alamos, New Mexico.
Dijkstra, Sandra. San Diego, California.
Epstein, Robert B. University of Oklahoma. Oklahoma City, Oklahoma.
Ezell, Edward Z. Smithsonian Institution. Washington, D.C.
Farber, Paul. Oregon State University. Corvallis, Oregon.
Feldberg, R. Tufts University. Medford, Massachusetts.
Graybar, Lloyd J. Eastern Kentucky University. Richmond, Kentucky.
Greb, G. Allen. University of California. San Diego, California.
Grimwood, James M. Houston, Texas.
Hacker, Sally L. Oregon State University. Corvallis, Oregon.
Haraway, Donna. University of California. Santa Cruz, California.
Hawthorne, Howard A. University of Utah. Salt Lake City, Utah.
Hewlett, Richard G. History Associates Incorporated. Washington, D.C.
Highland, Hugh. Navy Nuclear Test Personnel Review. Arlington, Virginia.
Holl, Jack M. Department of Energy. Washington, D.C.
Honicker, Clifford. University of Tennessee. Knoxville, Tennessee.
Hoos, Ida R. University of California. Berkeley, California.
Hufbauer, Karl. University of California. Irvine, California.

Hughes, Sally S. University of California. Berkeley, California.

Irving, Sutree. Corvallis, Oregon.

Jones, Scott. Kaman Tempo. Alexandria, Virginia.

Kaiser, Henry. Monmouth, Oregon.

Keller, Charles A. Oak Ridge Operations Office. Oak Ridge, Tennessee.

Kushner, Howard I. San Diego State University. San Diego, California.

Landsverk, Ole G. Glendale, California.

Loeffler, W. H. Navy Nuclear Test Personnel Review. Arlington, Virginia.

Lulejian, Norair M. Palos Verdes Estates, California.

McNeill, William H. University of Chicago. Chicago, Illinois.

Malik, John S. Los Alamos National Laboratory. Los Alamos, New Mexico.

Mardell, Jacqueline E. Northbrook, Illinois.

Marelli, Michael A. Nevada Operations Office. Las Vegas, Nevada.

Marlow, Sandra K. Mattapoisett, Massachusetts.

Martin, E. J. Kaman Tempo. Santa Barbara, California.

Melanson, Philip H. Southeastern Massachusetts University. North Dartmouth, Massachusetts.

Miller, David F. Nevada Operations Office. Las Vegas, Nevada.

Moore, Victor. Tillamook, Oregon.

Moroney, John D., III. Nevada Operations Office. Las Vegas, Nevada.

Mowery, H. B. Los Alamos National Laboratory. Los Alamos, New Mexico.

Nelson, Andrew G. JAYCOR. Alexandria, Virginia.

Noble, David F. Smithsonian Institution. Washington, D.C.

Nowack, Dorothy A. Nevada Operations Office. Las Vegas, Nevada.

Nutley, Richard V. Nevada Operations Office. Las Vegas, Nevada.

Page, Marshall, Jr. Nevada Operations Office. Las Vegas, Nevada.

Patterson, H. Wade Lawrence Livermore National Laboratory. Livermore, California.

Plummer, Grace M. Nevada Operations Office. Las Vegas, Nevada.

Reese, H. L. Defense Nuclear Agency. Washington, D.C.

Reingold, Nathan. Smithsonian Institution. Washington, D.C.

Remson, Alfred. East Norwalk, Connecticut.

Rivera, Tony A. Los Alamos National Laboratory. Los Alamos, New Mexico.

Robinette, C. Dennis. National Academy of Sciences. Washington, D.C.

Rosenberg, David A. University of Houston. Houston, Texas.

Ross, Donald M. Department of Energy. Germantown, Maryland.

Sacks, Karen. Silver Springs, Maryland.

Schneider, Joseph. Drake University. Des Moines, Iowa.

Spencer, J. Brooks. Oregon State University. Corvallis, Oregon.

Stine, Jeffrey K. Smithsonian Institution. Washington, D.C.

Stinson, Joe A. Defense Nuclear Agency. Kirtland Air Force Base, New Mexico.

Striegel, James F. JRB Associates. McLean, Virginia.

Szasz, Ferenc Morton. University of New Mexico. Albuquerque, New Mexico.

Thorne, Phillip D. Defense Nuclear Agency. Kirtland Air Force Base, New Mexico.

Walker, J. S. Nuclear Regulatory Commission. Washington, D.C.

Warnow, Joan N. American Institute of Physics. New York, New York.
Weart, Spencer R. American Institute of Physics. New York, New York
Weary, S. E. JAYCOR. Alexandria, Virginia.
Weiner, Charles. Massachusetts Institute of Technology. Cambridge, Massachusetts.
Wheeler, David L. Nevada Operations Office. Las Vegas, Nevada.

BIBLIOGRAPHY

Aebersold, Paul C. "July 16th Nuclear Explosion: Safety and Monitoring of Personnel." Report no. LAMS-616. Los Alamos Scientific Laboratory, 25 Jan. 1946.

Aebersold, Paul C., Louis H. Hempelmann, Jr., William H. Hinch, Joseph G. Hoffman, Wright H. Langham, and J. F. Tribby. "Chemistry and Metallurgy Health Handbook of Radioactive Materials." Los Alamos Scientific Laboratory, 17 Aug. 1945.

Allen, Kenneth D. A., ed. *Radiology in World War II.* Medical Department in World War II. Clinical Series. Washington: Office of the Surgeon General, U.S. Army, 1966.

Anderson, Herbert L. "Fermi, Szilard and Trinity." *Bulletin of the Atomic Scientists* 30 (Oct. 1974):40–47.

"Announced United States Nuclear Tests, July 1945 through December 1979." Report no. NVO-209. U.S. Department of Energy, Nevada Operations Office, Office of Public Relations. Las Vegas, Jan. 1981, and subsequent annual revisions.

Anton, Nicholas. "Radiation Counter Tubes and Their Operation." *Electronic Industries and Electronic Instrumentation* 2 (Feb. 1948):4–7.

Aston, Francis William. *Isotopes.* London: Edward Arnold, 1922.

Auxier, John A. *Ichiban: Radiation Dosimetry for the Survivors of Hiroshima and Nagasaki.* Oak Ridge, Tenn.: Technical Information Center, Energy Research and Development Administration, 1977.

Badash, Lawrence. "Chance Favors the Prepared Mind: Henri Becquerel and the Discovery of Radioactivity." *Archives Internationales d'Histoire des Sciences* 70 (1965):55–66.

————. "Radioactivity before the Curies." *American Journal of Physics* 33 (1965):128–135.

————. *Radioactivity in America: Growth and Decay of a Science.* Baltimore: Johns Hopkins University Press, 1979.

Badash, Lawrence, Joseph O. Hirschfelder, and Herbert P. Broida, eds. *Reminiscences of Los Alamos, 1943–1945.* Studies in the History of Modern Science. Vol. 5. Dordrecht, Holland: D. Reidel, 1980.

Bainbridge, Kenneth T. "A Foul and Awesome Display." *Bulletin of the Atomic Scientists* 31 (May 1975):40–46.

————. "Prelude to Trinity." *Bulletin of the Atomic Scientists* 31 (April 1975):42–46.

————. *Trinity.* Report no. LA-6300-H. Los Alamos, 1976. Declassified version of report no. LA-1012. Los Alamos Scientific Laboratory, 1946.

Barnaby, Frank. "The Continuing Body Count at Hiroshima and Nagasaki." *Bulletin of the Atomic Scientists* 33 (Dec. 1977):48–53.

Becker, Klaus. *Photographic Film Dosimetry: Principles and Methods of Quantitative Measurement of Radiation by Photographic Means.* Trans. K. S. Ankersmith. London and New York: Focal Press, 1966.

Bemis, Edwin A., Jr. "Survey Instruments and Pocket Dosimeters." In *Radiation Dosimetry*, ed. Gerald J. Hine and Gordon L. Brownell. New York: Academic Press, 1956. Pp. 454–503.

Berg, Samuel. "History of the First Survey on the Medical Effects of Radioactive Fall-Out." *Military Medicine* 124 (1959):782–785.

Berkhouse, L., S. E. Davis, F. R. Gladeck, J. H. Hallowell, C. B. Jones, E. J. Martin, F. W. McMullan, and M. J. Osborne. *Operation Crossroads, 1946.* Report no. DNA 6032F. Washington: Defense Nuclear Agency, 1 May 1984.

Bernstein, Barton J. "The Perils and Politics of Surrender: Ending the War with Japan and Avoiding the Third Atomic Bomb." *Pacific Historical Review* 46 (1977):1–28.

Bernstein, Jeremy. *Hans Bethe: Prophet of Energy.* New York: Basic Books, 1980.

Bhagavadgita. Various translations.

Blum, Theodore. "Osteomyelitis of the Mandible and Maxilla." *Journal of the American Dental Association* 11 (1924):802–805.

Blumberg, Stanley A., and Gwinn Owens. *Energy and Conflict: The Life and Times of Edward Teller.* New York: G. P. Putnam's Sons, 1976.

Boffey, Philip M. "Radiation Standards: Are the Right People Making Decisions?" *Science* 171 (1971):780–783.

Bond, Victor P., and J. W. Thiessen, eds. *Reevaluations of Dosimetric Factors: Hiroshima and Nagasaki.* Symposium. DOE Report no. CONF-810928. Oak Ridge, Tenn.: Technical Information Center, U.S. Department of Energy, 1982.

Boyer, Paul. *By the Bomb's Early Light: American Thought and Culture at the Dawn of the Atomic Age.* New York: Pantheon Books, 1985.

Bradley, David. *No Place to Hide.* Boston: Little, Brown, 1948.

Brecher, Ruth, and Edward Brecher. *The Rays: A History of Radiology in the United States and Canada.* Baltimore: Williams & Wilkins, 1969.

Brode, Bernice. "Tales of Los Alamos." In *Reminiscences of Los Alamos, 1943–1945*, ed. Lawrence Badash, Joseph O. Hirschfelder, and Herbert P. Broido. Studies in the History of Modern Science. Vol. 5. Dordrecht, Holland: D. Reidel, 1980. Pp. 133–159.

Brodsky, Allen B., ed. *CRC Handbook of Radiation Measurement and Protection.* Section A, vol. 1: *Physical Science and Engineering Data.* West Palm Beach, Fla.: CRC Press, 1978.

Brown, Anthony Cave, and Charles B. MacDonald, eds. *The Secret History of the Atomic Bomb.* New York: Dial Press/James Wade, 1977.

Bushong, Stewart C. "The Development of Current Radiation Protection Practices in Diagnostic Radiology." *CRC Critical Reviews in Radiological Sciences* 2 (1971):337–425.

Butow, Robert J. C. *Japan's Decision to Surrender.* Stanford: Stanford University Press; London: Geoffrey Cumberlege, Oxford University Press, 1954.

Cahn, Robert. "Behind the First A-Bomb." *Saturday Evening Post,* 16 July 1960, 16–17ff.

Cantril, Simeon T. "Biological Bases for Maximum Permissible Exposures." In *Industrial Medicine on the Plutonium Project: Survey and Collected Papers,* ed. Robert S. Stone. National Nuclear Energy Series. Div. IV, vol. 20. New York: McGraw-Hill, 1951. Pp. 36–74.

————. "Industrial Medical Program—Hanford Engineer Works." In *Industrial Medicine on the Plutonium Project: Survey and Collected Papers,* ed. Robert S. Stone. National Nuclear Energy Series. Div. IV, vol. 20. New York: McGraw-Hill, 1951. Pp. 289–307.

Cantril, Simeon T., and Herbert M. Parker. "Status of Health and Protection at the Hanford Engineer Works." In *Industrial Medicine on the Plutonium Project: Survey and Collected Papers,* ed. Robert S. Stone. National Nuclear Energy Series. Div. IV, vol. 20. New York: McGraw-Hill, 1951. Pp. 476–484.

————. "The Tolerance Dose." Health Division. Report no. CH-2812. Metallurgical Laboratory, 5 Jan. 1945.

Cantril, Simeon T., Ernest O. Wollan, and Kenneth S. Cole. "Health and Safety." Health Division. Report no. CH-376. Metallurgical Laboratory, 26 Nov. 1942.

Carlson, Elof Axel. *Genes, Radiation, and Society: The Life and Work of H. J. Muller.* Ithaca, N.Y.: Cornell University Press, 1981.

Case, James T. "The Early History of Radium Therapy and the American Radium Society." *American Journal of Roentgenology* 82 (1959):574–585.

Castle, William B., Katherine R. Drinker, and Cecil K. Drinker. "Necrosis of the Jaw in Workers Employed in Applying a Luminous Paint Containing Radium." *Journal of Industrial Hygiene* 7 (1925):317–382.

Chambers, Marjorie Bell. "Technically Sweet Los Alamos: The Development of a Federally Sponsored Scientific Community." Ph.D. dissertation. History. University of New Mexico, 1974.

Chorzempa, Mary André. "Ionizing Radiation and Its Chemical Effects: A Historical Study of Chemical Dosimetry (1902–1962)." Ph.D. dissertation. History of Science. Oregon State University, 1971.

Clark, George L., ed. *The Encyclopedia of X-Rays and Gamma Rays.* New York: Reinhold, 1963.

Clark, Ronald W. *The Birth of the Bomb.* New York: Horizon Press, 1961.

Cloutier, Roger J. "Florence Kelley and the Radium Dial Painters." *Health Physics* 39 (1980):711–716.

Coffin, Tris. "In the Light of Bikini." *Nation* 163 (5 Oct. 1946):370–371.

Coffinberry, A. S., and W. N. Miner, eds. *The Metal Plutonium*. Chicago: University of Chicago Press, 1961.

Collection of the Reports on the Investigation of the Atomic Bomb Casualties [in Japanese]. Science Council of Japan. Tokyo, 1951.

Collection of the Reports on the Investigation of the Atomic Bomb Casualties [in Japanese]. Science Council of Japan. 2 vols. Tokyo, 1953.

Collins, Donald L. "Pictures from the Past: Journeys into Health Physics in the Manhattan District and Other Diverse Places." In *Health Physics: A Backward Glance. Thirteen Original Papers on the History of Radiation Protection*, ed. Ronald L. Kathren and Paul L. Ziemer. New York: Pergamon Press, 1980. Pp. 37–71.

Collins, Donald L., ed. "Operations Crossroads Reports to Rad Safe Instrument Division." Radiological Safety Section, Joint Task Force One, n.d.

Commander, Joint Task Force One. "Aerological Report on Operation Crossroads." Report no. OPNAV-JTF-P1001. Joint Task Force One, n.d.

————. Operation Plan no. 1-46. Joint Task Force One, 15 April 1946.

————. "Report on Atomic Bomb Tests Able and Baker (Operation Crossroads) Conducted at Bikini Atoll, Marshall Islands, on 1 July 1946 and 25 July 1946." Joint Task Force One, 15 Nov. 1946.

Compton, Arthur H. *Atomic Quest: A Personal Narrative*. New York: Oxford University Press, 1956.

Coolidge, W. D., and E. E. Charlton, "Roentgen-Ray Tubes." *Radiology* 45 (1945):449–466.

Cooney, James P. "Medical Effects of Atomic Explosion." In *Radiological Defense. Vol. 3: A Series of Indoctrination Lectures on Atomic Explosion, with Medical Aspects*. Armed Forces Special Weapons Project. Washington, n.d. Pp. 30–36.

Cope, Zachary, ed. *Sidelights on the History of Medicine*. London: Butterworth, 1957.

Cottrell, Leonard S., Jr., and Sylvia Eberhart. *American Opinion on World Affairs in the Atomic Age*. Princeton: Princeton University Press, 1948.

Cowie, Dean B., and Leonard A. Scheele. "A Survey of Radiation Protection in Hospitals." *Journal of the National Cancer Institute* 1 (1941):767–787. Repr. in *Health Physicis* 38 (1980):929–947.

Craven, Wesley Frank, and James Lea Cate, eds. *The Army Air Forces in World War II*. Vol. 5: *The Pacific: Matterhorn to Nagasaki, June 1944 to August 1945*. Chicago and London: University of Chicago Press, 1953.

Craver, L. F. "Tolerance to Whole-Body Irradiation of Patients with Advanced Cancer." In *Industrial Medicine on the Plutonium Project: Survey and Collected Papers*, ed. Robert S. Stone. National Nuclear Energy Series. Div. IV, vol. 20. New York: McGraw-Hill, 1951. Pp. 485–498.

Cumberledge, A. A. "Aerological Aspects of the Bikini Bomb Tests." *Scientific Monthly* 64 (Feb. 1947): 135–147.

Defense Nuclear Agency, Public Affairs Office. "Hiroshima and Nagasaki Occupation Forces." Fact Sheet. Washington, 6 Aug. 1980.

————. "Nuclear Test Personnel Review (NTPR)." Fact Sheet. Washington, 1 March 1982.

————. "Operation CROSSROADS." Fact Sheet. 5 April 1984.

Director of Ship Material. "Radiological Decontamination of Target and Non-Target Vessels." By J. J. Fee. 3 vols. Technical Reports XRD-185, -186, -187. Joint Task Force One, 1946.

————. "Technical Inspection Report: Test A and Test B." By F. X. Forest. Bureau of Ships Group Final Report no. XRD-2, vol. 1. Joint Task Force One, 16 Dec. 1946.

Dowdy, Andrew H. "University of Rochester Project Foreword." In *Pharmacology and Toxicology of Uranium Compounds, with a Section on the Pharmacology and Toxicology of Fluorine and Hydrogen Fluoride,* ed. Carl Voegtlin and Harold C. Hodge. National Nuclear Energy Series. Div. VI, vol. 1, part 1. New York: McGraw-Hill, 1949–1953. Pp. xi-xii.

Draeger, R. H., and Shields Warren. "Medicine at the Crossroads." *U.S. Naval Medical Bulletin* 47 (1947):219–225.

Dunham, Charles L. "Symposium on Medicine and Atomic Warfare: Medical Aspects of Atomic Warfare." *Connecticut State Medical Journal* 15 (1951):1039–1047.

Dunham, Charles L., Eugene P. Cronkite, George V. LeRoy, and Shields Warren. "Atomic Bomb Injury: Radiation." *Journal of the American Medical Association* 147 (1951):50–54.

Dupree, A. Hunter. "The Great *Instauration* of 1940: The Organization of Scientific Research for War." In *The Twentieth-Century Sciences: Studies in the Biography of Ideas,* ed. Gerald Holton. New York: W. W. Norton, 1972. Pp. 443–467.

Easlea, Brian. *Fathering the Unthinkable: Masculinity, Scientists and the Nuclear Arms Race.* London: Pluto Press, 1983.

Eberhart, Sylvia. "How the American People Feel about the Atomic Bomb." *Bulletin of the Atomic Scientists* 3 (June 1947):146–149f.

Eisenbud, Merril. "Early Occupational Exposure Experience with Uranium Processing." In *Proceedings of the Conference on Occupational Health Experience with Uranium.* Report no. 93 UC 41. Energy Research and Development Administration. Washington, 1976.

————. *Environmental Radioactivity.* 2d ed. New York: Academic Press, 1973.

Eisenbud, Merril, and J. A. Quigley. "Industrial Hygiene of Uranium Processing." *A.M.A. Archives of Industrial Health* 14 (July 1946):12-22.

Erskine, Hazel Gaudet. "The Polls: Atomic Weapons and Nuclear Energy." *Public Opinion Quarterly* 27 (1963):155–190.

Evans, Robley D. "Inception of Standards for Internal Emitters, Radon and Radium." *Health Physics* 41 (1981):437-448.

————. "Origin of Standards for Internal Emitters." In *Health Physics: A Backward Glance. Thirteen Original Papers on the History of Radiation Protection,* ed. Ronald L. Kathren and Paul L. Ziemer. New York: Pergamon Press, 1980. Pp. 141–157.

————. "Protection of Radium Dial Workers and Radiologists from Injury by Radium." *Journal of Industrial Hygiene and Toxicology* 25 (1943):253-269.

————. "Radium in Man." *Health Physics* 27 (1974):497-510.

————. "Radium Poisoning: A Review of Present Knowledge." *American Journal of Public Health* 23 (1933):1017-1023.

Evans, Robley D., and Clark Goodman. "Determination of the Thoron Content of Air and Its Bearing on Lung Cancer Hazards in Industry." *Journal of Industrial Hygiene and Toxicology* 22 (1940):89-99. Repr. in *Health Physics* 38 (1980):919-928.

Failla, Gioacchino. "Biological Effects of Ionizing Radiation." *Journal of Applied Physics* 12 (1941):279-295.

————. "Ionization Measurements." *American Journal of Roentgenology* 10 (1923):48-56. Repr. in *Health Physics* 38 (1980):889-897.

Farrell, Thomas F. "Memorandum to Major General L. R. Groves: Subject: Report on Overseas Operations—Atomic Bomb," 27 September 1945. In *The Secret History of the Atomic Bomb*, ed. Anthony Cave Brown and Charles B. MacDonald. New York: Dial Press, 1977. Pp. 529-538.

Fermi, Laura. *Atoms in the Family: My Life with Enrico Fermi*. Chicago: University of Chicago Press, 1954.

Fine, Lenore, and Jesse A. Remington. *The Corps of Engineers: Construction in the United States*. United States Army in World War II: The Technical Services. Washington: Office of the Chief of Military History, 1972.

Fitch, Val. "The View from the Bottom." *Bulletin of the Atomic Scientists* 31 (Feb. 1975):43-46.

Friedell, Hymer L. "Early Atomic Bomb Casualty Commission Perceptions and Planning." In *Reevaluations of Dosimetric Factors: Hiroshima and Nagasaki*, ed. Victor P. Bond and J. W. Thiessen. DOE Report no. CONF-810928. Oak Ridge, Tenn.: Technical Information Center, 1982. Pp. 1-5.

Frisch, Otto R. "The Los Alamos Experience." *New Scientist* 83 (19 July 1979):186-188.

————. "Somebody Turned the Sun on with a Switch." *Bulletin of the Atomic Scientists* 30 (April 1974):12-18.

————. *What Little I Remember*. Cambridge: Cambridge University Press, 1979.

Fuchs, Arthur W. "Evolution of Roentgen Film." *American Journal of Roentgenology* 75 (1956):30-48.

Furter, William F., ed. *History of Chemical Engineering*. Washington: American Chemical Society, 1980.

Gillispie, Charles Coulston, ed. *Dictionary of Scientific Biography*. Vol. 11. New York: Charles Scribner's Sons, 1975.

Glasser, Otto. "Technical Development of Radiology." *American Journal of Roentgenology* 75 (1956):7-13.

————. *Wilhelm Conrad Röntgen and the Early History of Röntgen Rays*. Springfield, Ill.: Charles C. Thomas, 1934.

Glasstone, Samuel, ed. *The Effects of Atomic Weapons*. Prepared for and in cooperation with U.S. Department of Defense and U.S. Atomic Energy

Commission under direction of Los Alamos Scientific Laboratory. Washington: Government Printing Office, 1950, and later editions.

———. *Sourcebook on Atomic Energy.* New York: D. Van Nostrand, 1950.

Gofman, John W. *Radiation and Human Health.* San Francisco: Sierra Club Books, 1981.

Goodchild, Peter. *J. Robert Oppenheimer: Shatterer of Worlds.* Boston: Houghton Mifflin, 1981.

Goudsmit, S. A. *Alsos.* New York: Henry Schumann, 1947.

Gowing, Margaret. *Britain and Atomic Energy, 1939–1945.* London: Macmillan; New York: St. Martin's Press, 1964.

Graybar, Lloyd J. "Bikini Revisited." *Military Affairs* 44 (1980):118–123.

Graybar, Lloyd J., and Ruth Flint Graybar. "America Faces the Atomic Age, 1946." *Air University Review* 35 (Jan.–Feb. 1984):68–77.

Greinacher, H. "The Evolution of Particle Counters." *Endeavour* 13 (Oct. 1954):190–197.

Grigg, E. R. N. *The Trail of the Invisible Light: From X-Strahlen to Radio(bio)logy.* Springfield, Ill.: Charles C. Thomas, 1965.

Grosch, Daniel S., and Larry E. Hopwood. *Biological Effects of Radiations.* 2d ed. New York: Academic Press, 1979.

Groueff, Stephane. *Manhattan Project: The Untold Story of the Making of the Atomic Bomb.* Boston and Toronto: Little, Brown, 1967.

Groves, Leslie R. "The A-Bomb Program." In *Science, Technology, and Management,* ed. Fremont E. Kast and James E. Rosenzweig. New York: McGraw-Hill, 1963. Pp. 31–40.

———. *Now It Can Be Told: The Story of the Manhattan Project.* New York and Evanston: Harper & Row, 1962.

———. "Some Recollections of July 16, 1945" *Bulletin of the Atomic Scientists* 26 (June 1970): 21–27.

Hamilton, Joseph G. "The Metabolism of the Fission Products and the Heaviest Elements." *Radiology* 49 (1947):325–343.

Hammond, Alexander. Review of *No Place to Hide, 1946/1984* by David Bradley. *Bulletin of the Atomic Scientists* 39 (Nov. 1983):37–39.

Harris, Benedict R., and Marvin A. Stevens. "Experiences at Nagasaki, Japan." *Connecticut State Medical Journal* 9 (1945):913–917.

Hawkins, David. "Manhattan District History, Project Y, the Los Alamos Project." Vol. 1: "Inception until August 1945." Report no. LAMS-2532 (Vol. I). Los Alamos Scientific Laboratory, completed Aug. 1946, issued Dec. 1961.

———. "Towards Trinity." In *Project Y: The Los Alamos Story.* History of Modern Physics, 1800–1950. Vol. 2. Los Angeles and San Francisco: Tomash, 1983. Part I. Published version of report above.

Hawthorne, Howard A., ed. *Compilation of Local Fallout Data from Test Detonations, 1945–1962. Extracted from DASA 1251.* 2 vols. Report no. DASA 1251(EX). Defense Nuclear Agency. Washington, May 1979.

Hazard Control Committee. "Product [Plutonium] Hazard Control Regulations and General Safety Precautions." Chemistry Division. Report no. CN-2408. Metallurgical Laboratory, 20 Nov. 1944.

————. "Rules for Fission Product Hazard Control and Recommendations for Operations with Fission Products." Health Division. Report no. CH-2493. Metallurgical Laboratory, 19 Dec. 1944.

Health Division. "Health Division Program." Report no. CH-632. Metallurgical Laboratory, 10 May 1943.

————. "Site X Operating Manual: Health Hazards." Report no. CH-732. Metallurgical Laboratory, May 1943.

Hempelmann, Louis H., Jr. "Hazards of Trinity Experiment, Section I." Los Alamos Scientific Laboratory, 12 April 1945.

————. "History of the Health Group (A-6) (March 1943–November 1945)." Los Alamos Scientific Laboratory, April 1946.

————. "History of the Preparation of the Medical Group for Trinity Test II." Los Alamos Scientific Laboratory, n.d.

————. "Preparation and Operational Plan of Medical Group (TR-7) for Nuclear Explosion 16 July 1945." Report no. LA-631. Los Alamos Scientific Laboratory, 13 June 1947.

Hempelmann, Louis H., Jr., comp., "Nuclear Explosion 16 July 1945: Health Physics Report on Radioactive Contamination throughout New Mexico, Part B: Biological Effects." Los Alamos Scientific Laboratory, ca. April 1946.

Hempelmann, Louis H., Jr., Hermann Lisco, and Joseph G. Hoffman. "The Acute Radiation Syndrome: A Study of Nine Cases and a Review of the Problem." *Annals of Internal Medicine* 36 (1952):279–500.

Henshaw, Paul S. "Biological Significance of the Tolerance Dose in X-Ray and Radium Protection." *Journal of the National Cancer Institute* 1 (1941):789–805.

Herken, Gregg. *The Winning Weapon: The Atomic Bomb and the Cold War, 1945–1950.* New York: Alfred A. Knopf, 1980.

Hersey, John. *Hiroshima.* New York: Alfred A. Knopf, 1946.

Hewlett, Richard G., and Oscar E. Anderson, Jr. *A History of the United States Atomic Energy Commission.* Vol. 1: *The New World, 1939/1946.* University Park: Pennsylvania State University Press, 1962.

Hewlett, Richard G., and Francis Duncan. *A History of the United States Atomic Energy Commission.* Vol. 2: *Atomic Shield, 1947/1952.* University Park: Pennsylvania State University Press, 1969.

Hine, Gerald J., and Gordon L. Brownell, eds. *Radiation Dosimetry.* New York: Academic Press, 1956.

Hines, Neal O. *Proving Ground: An Account of the Radiobiological Studies in the Pacific, 1946–1961.* Seattle: University of Washington Press, 1962.

Hiroshima and Nagasaki: The Physical, Medical, and Social Effects of the Atomic Bombing. Committee for the Compilation of Materials on Damage Caused by the Atomic Bombs in Hiroshima and Nagasaki. Trans. Eisei Ishikawa and David L. Swain. New York: Basic Books, 1981.

Hirschfelder, Joseph O. "The Scientific and Technological Miracle at Los Alamos." In *Reminiscences of Los Alamos, 1943–1945*, ed. Lawrence Badash, Joseph O. Hirschfelder, and Herbert P. Broida. Studies in the History of Modern Science. Vol. 5. Dordrecht, Holland: D. Reidel, 1980. Pp. 67–88.

Hodge, Harold C. "Historical Foreword." In *Pharmacology and Toxicology of Uranium Compounds, with a Section on the Pharmacology and Toxicology of Fluorine and Hydrogen Fluoride*, ed. Carl Voegtlin and Harold C. Hodge. National Nuclear Energy Series. Div. VI, vol. 1, part 1. New York: McGraw-Hill, 1949–1953. Pp. 1–14.

Hodge, Harold C., J. Newell Stannard, and J. B. Hursh, eds. *Uranium, Plutonium, Transplutonic Elements*. Handbook of Experimental Pharmacology. Vol. 36. New York: Springer-Verlag, 1972.

Hoffman, Frederic de. "Pure Science in the Service of Wartime Technology." *Bulletin of the Atomic Scientists* 31 (Jan. 1975): 41–44.

Hoffman, Frederick L. "Radium (Mesothorium) Necrosis." *Journal of the American Medical Association* 85 (1925):961-965.

Hoffman, Joseph G. "Nuclear Explosion July 16 1945: Health Physics Report on Radioactive Contamination throughout New Mexico Following the Nuclear Explosion, Part A: Physics." Report no. LA-626. Los Alamos Scientific Laboratory, 20 Feb. 1947.

Holton, Gerald, ed. *The Twentieth-Century Sciences: Studies in the Biography of Ideas*. New York: W. W. Norton, 1972.

Howland, Joe W., and Stafford L. Warren. "The Effects of the Atomic Bomb Irradiation on the Japanese." *Advances in Biological and Medical Physics* 1 (1948):387–408.

Irving, David. *The German Atomic Bomb: The History of Nuclear Research in Nazi Germany*. New York: Simon & Schuster, 1967.

Jacobson, Leon O., and Edna K. Marks. "Clinical Laboratory Examination of Plutonium Project Personnel." In *Industrial Medicine on the Plutonium Project: Survey and Collected Papers*, ed. Robert S. Stone. National Nuclear Energy Series. Div. IV, vol. 20. New York: McGraw-Hill, 1951. Pp. 113–139.

———. "The Hematological Effects of Ionizing Radiations in the Tolerance Range." *Radiology* 49 (1947):286–298.

Jacobson, Leon O., Edna K. Marks, and Egon Lorenz. "Hematological Effects of Ionizing Radiation." In *Industrial Medicine on the Plutonium Project: Survey and Collected Papers*, ed. Robert S. Stone. National Nuclear Energy Series. Div. IV, vol. 20. New York: McGraw-Hill, 1951. Pp. 140–196.

Jauncey, G. E. M. "The Birth and Early Infancy of X-Rays." *American Journal of Physics* 13 (1945):362–379.

Jesse, William P. "The Role of Instruments in the Atomic Bomb Project." *Chemical Engineering News* 24 (1946):2906–2909.

Johnson, Charles W., and Charles O. Jackson. *City behind a Fence: Oak Ridge, Tennessee, 1942–1946*. Knoxville: University of Tennessee Press, 1981.

Joint Chiefs of Staff Evaluation Board for Atomic Bomb Tests. "Preliminary Report Following the First Atomic Bomb Test." Operation Crossroads, 11 July 1946.

————. "Preliminary Report Following the Second Atomic Bomb Test." Operation Crossroads, 30 July 1946.

Joint Crossroads Committee. "Radiological Safety." N.d. Published as *Radiological Defense*. Vol. 1. Armed Forces Special Weapons Project. Washington, 1948.

Jungk, Robert. *Brighter than a Thousand Suns: A Personal History of the Atomic Scientists*. Trans. James Cleugh. New York: Harcourt, Brace, 1958.

Kast, Fremont E., and James E. Rosenzweig, eds. *Science, Technology, and Management*. New York: McGraw-Hill, 1963.

Kathren, Ronald L. "Before Transistors, IC's, and All Those Other Good Things: The First Fifty Years of Radiation Instrumentation." In *Health Physics: A Backward Glance. Thirteen Original Papers on the History of Radiation Protection*, ed. Ronald L. Kathren and Paul L. Ziemer. New York: Pergamon Press, 1980. Pp. 73–93.

————. "Early X-Ray Protection in the United States." *Health Physics* 8 (1962):503–511.

————. "Historical Development of Radiation Measurement and Protection." In *CRC Handbook of Radiation Measurement and Protection*. Section A, vol. 1: *Physical Science and Engineering Data*, ed. Allen B. Brodsky. West Palm Beach, Fla.: CRC Press, 1978. Pp. 13–52.

————. "William H. Rollins (1852–1929): X-Ray Protection Pioneer." *Journal of the History of Medicine and Allied Sciences* 19 (1964):287–294.

Kathren, Ronald L., and Paul L. Ziemer, eds. *Health Physics: A Backward Glance. Thirteen Original Papers on the History of Radiation Protection*. New York: Pergamon Press, 1980.

Kauffman, Draper L. "Report of the Radiological Safety School, Joint Task Force One, Navy Department, Washington, 9 September 1946 through 9 October 1946." 14 Oct. 1946.

Kevles, Daniel J. *The Physicists: The History of a Scientific Community in Modern America*. New York: Alfred A. Knopf, 1977.

Kirkpatrick, S. D. "A-Bomb Tests: As Viewed by the Editor." *Chemical Engineering* 53 (Aug. 1946):94–96ff.

Kiste, Robert C. *The Bikinians: A Study in Forced Migration*. Kiste-Ogan Social Change Series in Anthropology. Menlo Park, Cal.: Cummings, 1974.

Kistiakowsky, George B. "Reminiscences of Wartime Los Alamos." In *Reminiscences of Los Alamos, 1943–1945*, ed. Lawrence Badash, Joseph O. Hirschfelder, and Herbert P. Broida. Studies in the History of Modern Science. Vol. 5. Dordrecht, Holland: D. Reidel, 1980. Pp. 49–65.

Knebel, Fletcher, and Charles W. Bailey II. *No High Ground*. New York: Harper & Row, 1960. Repr. New York: Bantam Books, 1961.

Knight, Jerry. "U.S. Hunts A-Bomb Project Debris." *Los Angeles Times*, 20 Jan. 1981. Section IV, p. 13.

Kunetka, James W. *City of Fire: Los Alamos and the Birth of the Atomic Age, 1943–1945.* Englewood Cliffs, N.J.: Prentice-Hall, 1978.

Lacassagne, A. "Historical Outline of the Initial Studies on the Use of Polonium in Biology." *British Journal of Radiology* 44 (1971):546–548.

Lamont, Lansing. *Day of Trinity.* New York: Atheneum, 1965.

Landa, Edward R. "The First Nuclear Industry." *Scientific American* 247 (Nov. 1982):180ff.

Lang, Daniel. "A Fine Moral Point." *New Yorker,* Jan. 1946. Repr. in Lang, *From Hiroshima to the Moon.* New York: Dell, 1961. Pp. 51–65.

———. *From Hiroshima to the Moon.* New York: Dell, 1961.

———. "A Most Valuable Accident." *New Yorker,* 2 May 1959, 49ff. Repr. in Lang, *From Hiroshima to the Moon.* New York: Dell, 1961. Pp. 418–445.

Langham, Wright H., Samuel H. Bassett, Payne S. Harris, and Robert E. Carter. "Distribution and Excretion of Plutonium Administered Intravenously to Man." Report no. LA-1151. Los Alamos Scientific Laboratory, 20 Sept. 1950. Published in *Health Physics* 38 (1980):1031–1060.

Langham, Wright H., and John W. Healy. "Maximum Permissible Body Burdens and Concentrations of Plutonium: Biological Basis and History of Development." In *Uranium, Plutonium, Transplutonic Elements,* ed. Harold C. Hodge, J. Newell Stannard, and J. B. Hursh. Handbook of Experimental Pharmacology. Vol. 36. New York: Springer-Verlag, 1973. Pp. 569–592.

Lapp, Ralph E. "Survey of Nucleonics Instrumentation Industry." *Nucleonics* 4 (May 1949):100–104.

Larson, Kermit H. "Continental Close-in Fallout: Its History, Measurement and Characteristics." In *Radioecology,* ed. Vincent Schultz and Alfred W. Klement. New York: Reinhold; Washington: American Institute of Biological Sciences, 1961. Pp. 19–25.

Laurence, William L. *Dawn over Zero: The Story of the Atomic Bomb.* 2d ed. New York: Alfred A. Knopf, 1946. Repr. Westport, Conn.: Greenwood Press, 1972.

———. *Men and Atoms: The Discovery, the Uses and the Future of Atomic Energy.* New York: Simon & Schuster, 1959.

Libby, Leona Marshall. *The Uranium People.* New York: Crane Russak and Charles Scribner's Sons, 1979.

Loeb, Paul. *Nuclear Culture: Living and Working in the World's Largest Atomic Complex.* New York: Coward, McCann & Geoghegan, 1982.

Loewe, W. E., and E. Mendelsohn. "Revised Dose Estimates at Hiroshima and Nagasaki." *Health Physics* 41 (1981):663–666.

Lorenz, Egon, and Walter E. Heston. "Effects of Long-Continued Total-Body Gamma Irradiation on Mice, Guinea Pigs, and Rabbits. 1. Preliminary Experiments." In *Biological Effects of External X and Gamma Radiation,* ed. Raymond E. Zirkle. National Nuclear Energy Series. Div. IV, vol. 22B. New York: McGraw-Hill, 1954. Pp. 1–11.

Lowrance, William W. *Of Acceptable Risk: Science and the Determination of Safety.* Los Altos, Cal.: William Kaufmann, 1976.

Lundie, A. "Some Medical Aspects of Atomic Warfare." *Journal of the Royal Army Medical Corps* 94 (1950):246–258.

Lyon, George M. "Radiological Defense." In *Radiological Defense*. Vol. 3: *A Series of Indoctrination Lectures on Atomic Explosion, with Medical Aspects*. Armed Forces Special Weapons Project. Washington, n.d. Pp. 97–109.

Maag, Carl, and Steve Rohrer. *Project Trinity, 1945–1946*. DNA report no. 6028F. Washington: Defense Nuclear Agency, 15 Dec. 1982.

McDaniel, Boyce. "A Physicist at Los Alamos." *Bulletin of the Atomic Scientists* 30 (Dec. 1974):39–43.

McRaney, W., and J. McGahan. "Radiation Dose Reconstruction, U.S. Occupation Forces in Hiroshima and Nagasaki, Japan, 1945–1946." Report no. 5512F. Defense Nuclear Agency, Washington, 6 Aug. 1980.

Makhijani, Arjun, and David Albright. "Irradiation of Personnel during Operation Crossroads: An Evaluation Based on Official Documents." International Radiation Research and Training Institute. Washington, May 1983.

Maki, Hiroshi, Isamu Nagai, Atsumu Okada, Mitsugu Usagawa, and Kimiko Ono. *The Atomic Bomb Casualty Commission, 1947–1975: A General Report on the ABCC-JNIH Joint Research Program*. Report no. TID-28719. National Academy of Sciences–National Research Council and Japanese National Institute of Health of the Ministry of Health and Welfare. Washington and Tokyo, 1978.

Manhattan Project Atomic Bomb Investigation Group. "The Atomic Bombings of Hiroshima and Nagasaki." Manhattan Engineer District, 30 June 1946.

"The Manhattan Project Atomic-Bomb Investigating Group." Washington, ca. 1946. Published in *The Secret History of the Atomic Bomb*, ed. Anthony Cave Brown and Charles B. MacDonald. New York: Dial Press, 1977. Pp. 552-582.

Mann, Wilfrid B., and S. B. Garfinkel. *Radioactivity and Its Measurement*. Princeton, N.J.: D. Van Nostrand, 1966.

Markwith, Carl. "Farewell to Bikini." *National Geographic* 90 (July 1946):97–116.

Martland, Harrison S. "Occupational Poisoning in Manufacture of Luminous Watch Dials: General Review of Hazard Caused by Ingestion of Luminous Paint, with Especial Reference to the New Jersey Cases." *Journal of the American Medical Association* 92 (1929):466–473.

———. "The Occurrence of Malignancy in Radioactive Persons." *American Journal of Cancer* 15 (1931):2435-2516.

Martland, Harrison S., Philip Conlon, and Joseph P. Knef. "Some Unrecognized Dangers in the Use and Handling of Radioactive Substances, with Especial Reference to the Storage of Insoluble Products of Radium and Mesothorium in the Reticulo-Endothelial System." *Journal of the American Medical Association* 85 (1925):1769–1776.

Melanson, Philip H. "The Human Guinea Pigs at Bikini." *Nation*, 9–16 July 1983, 1ff.

Millis, Walter. *Arms and Men: A Study in American Military History*. New York: G. P. Putnam's Sons, 1956. Repr. New York: Mentor Books, n.d.

Moll, Kenneth L. "Operation Crossroads: The Bikini A-Bomb Tests, 1946." *Air Force Magazine* 54 (July 1971):62–69.

Morgan, Karl Z. "Final Report of the Alpha Beta Gamma Survey Section." Radiological Safety Section, Joint Task Force One, 6 Aug. 1946.

———. "Instrumentation in the Field of Health Physics." *Proceedings of the I.R.E.* 37 (Jan. 1949):74–82.

Muller, Herbert J. "Artifical Transmutation of the Gene." *Science* 66 (1927):84–87.

Mutscheller, Arthur. "Physical Standards of Protection against Roentgen-Ray Dangers." *American Journal of Radiology* 13 (1925):65–70.

National Bureau of Standards. *Radium Protection for Amounts up to 300 Milligrams.* NBS Handbook 18. Washington, 17 May 1934.

———. *Safe Handling of Radioactive Luminous Compound.* NBS Handbook 27. Washington, 2 May 1941.

———. *X-Ray Protection.* NBS Handbook 15. Washington, 16 May 1931.

National Research Council. "Genetic Effects of Atomic Bombs in Hiroshima and Nagasaki." Committee on Atomic Casualties, Genetics Conference. *Science* 106 (1947):331–333.

———. "Multiple Myeloma among Hiroshima/Nagasaki Veterans." Washington, 15 July 1983.

———. "Report of Panel on Feasibility and Desirability of Performing Epidemiological Studies on U.S. Veterans of Hiroshima and Nagasaki." Washington, 21 Aug. 1981.

Neel, James V. "Genetics Conference." University of Michigan, 10–11 July 1953.

Newell, Robert R. "Report of Medic[o]-Legal Board." Radiological Safety Section, Joint Task Force One, 19 Aug. 1946.

Nickson, J. J. "Protective Measures for Personnel." In *Industrial Medicine on the Plutonium Project: Survey and Collected Papers,* ed. Robert S. Stone. National Nuclear Energy Series. Div. IV, vol. 20. New York: McGraw-Hill, 1951. Pp. 75–112.

Nolan, James F. "Medical Hazards of TR #2." Los Alamos Scientific Laboratory, 20 June 1945.

Norwood, W. Daggett. "Study of Chemical Hazards in Extraction Plant." Metallurgical Laboratory, 24 April 1943.

Office of Technology Assessment. "Review of the Report, 'Multiple Myeloma among Hiroshima/Nagasaki Veterans.'" Health Program, 17 Dec. 1983.

Okada, S., H. B. Hamilton, N. Egami, S. Okajima, W. J. Russell, and K. Takeshita, eds. "A Review of Thirty Years Study of Hiroshima and Nagasaki Atomic Bomb Survivors." *Journal of Radiation Research* 16 (1975), Supplement.

Okajima, Shunzo, Kenji Takeshita, Shigetoshi Antoku, Toshio Shiomi, Walter J. Russell, Shoichiro Fujita, Haruma Yoshinaga, Shotaro Neriishi, Sadahisa Kawamoto, and Toshiyuki Norimura. "Radioactive Fallout Effects of the Nagasaki Atomic Bomb." *Health Physics* 34 (1978):621–633.

"Operation Crossroads." *National Geographic* 91 (April 1947). Separately paginated photographs and paintings.

Operation Crossroads: The Official Pictorial Record. Office of the Historian, Joint Task Force One. New York: Wm. H. Wise, 1946.

Oppenheimer, J. Robert. *Robert Oppenheimer: Letters and Recollections,* ed. Alice Kimball Smith and Charles Weiner. Cambridge, Mass., and London: Harvard University Press, 1980.

Orient, Jane M. "A Critique of the Statistical Methodology in *Radiation and Human Health* by John W. Gofman." *Health Physics* 45 (1983):823–827.

Oughterson, Ashley W., and Shields Warren, eds. *Medical Effects of Atomic Bombs: The Report of the Joint Commission for the Effects of the Atomic Bomb in Japan.* 6 vols. Atomic Energy Commission. Washington, 1951.

————. *Medical Effects of the Atomic Bomb in Japan.* National Nuclear Energy Series. Div. VIII, vol. 8. New York: McGraw-Hill, 1956.

Oughterson, Ashley W., Shields Warren, Averill A. Liebow, George V. LeRoy, E. Cuyler Hammond, Henry L. Barnett, Jack D. Rosenbaum, and B. Aubrey Schneider. "Medical Report of the Joint Commission for the Investigation of the Effects of the Atomic Bomb in Japan." Sept. 1946.

Pace, Nello, and Robert E. Smith. "Measurement of the Residual Radiation Intensity at the Hiroshima and Nagasaki Atomic Bomb Sites." Report no. 160A. Naval Medical Research Institute, 16 April 1946.

Pardue, L. A., N. Goldstein, and Ernest O. Wollan. "Photographic Film as a Pocket Radiation Dosimeter." Health Physics Section, Health Division. Report no. CH-1553. Metallurgical Laboratory, 8 April 1944.

Parker, Herbert M. "Health-Physics, Instrumentation, and Radiation Protection." *Advances in Biological and Medical Physics* 1 (1948):223–285. Repr. in *Health Physics* 38 (1980):957–996.

————. "Protection Programs of the Plutonium Project." *Health Physics* 41 (1981):571-576.

————. "Some Background Information on the Development of Dose Units." Report no. CRUSP-1. Commission on Radiologic Units, Standards, and Protection. American College of Radiology, Nov. 1955.

————. "Tentative Dose Units for Mixed Radiations." *Radiology* 54 (1950):257–261. Repr. in *Health Physics* 38 (1980):1021–1024.

Parker, Herbert M., and William C. Roesch. "Units, Radiation: Historical Development." In *The Encyclopedia of X-Rays and Gamma Rays,* ed. George L. Clark. New York: Reinhold, 1963. Pp. 1102–1107.

Pash, Boris T. *The Alsos Mission.* New York: Charter Books, 1980. Repr. of 1969 ed.

Peterson, Arthur V. "Peppermint." Washington, ca. 1946. Published in *The Secret History of the Atomic Bomb,* ed. Anthony Cave Brown and Charles B. MacDonald. New York: Dial Press, 1977. Pp. 234–238.

"The Physical and Medical Effects of the Hiroshima and Nagasaki Bombs." Report of the Natural Science Group organized by the International Peace Bureau, Geneva. *Bulletin of the Atomic Scientists* 33 (Dec. 1977):54–56.

"The Plutonium Project." Symposium. *Radiology* 49 (1947):269–365.

Polednak, Anthony P. "Bone Cancer among Female Radium Dial Workers: Latency Periods and Incidence Rates by Time after Exposure." *Journal of the National Cancer Institute* 60 (1978):77–82.

Polednak, Anthony P., Andrew F. Stehney, and R. E. Rowland. "Mortality among Women First Employed before 1930 in the U.S. Radium Dial-Painting Industry." *Journal of Epidemiology* 107 (1978):179–195.

President's Evaluation Commission. "Preliminary Report on First Bikini Atom Bomb Test." Operation Crossroads, 11 July 1946.

————. "Preliminary Report on the Second Bikini Atom Bomb Test." Operation Crossroads, 29 July 1946.

Project Y: The Los Alamos Story. The History of Modern Physics, 1800–1950. Vol. 2. Los Angeles and San Francisco: Tomash, 1983.

Pursell, Carroll W., Jr. "Science Agencies in World War II: The OSRD and Its Challengers." In *The Sciences in the American Context: New Perspectives*, ed. Nathan Reingold. Washington: Smithsonian Institution Press, 1979. Pp. 359–378.

Quimby, Edith H. "The Background of Radium Therapy in the United States, 1906–1956." *American Journal of Roentgenology* 75 (1956):443–450.

————. "The History of Dosimetry in Roentgen Therapy." *American Journal of Roentgenology* 54 (1945):688–703.

————. "Radium Protection." *Journal of Applied Physics* 10 (1939):604–608.

Radiological Defense. Vol. 1. Armed Forces Special Weapons Project. Washington, 1948.

Radiological Defense. Vol. 2: *The Prinicples of Military Defense against Atomic Weapons.* Armed Forces Special Weapons Project. Washington, Nov. 1951.

Radiological Defense. Vol. 3: *A Series of Indoctrination Lectures on Atomic Explosion, with Medical Aspects.* Armed Forces Special Weapons Project. Washington, n.d.

Radiological Safety Section. "Nuclear Radiation Effects in Tests Able and Baker, Preliminary Report of." Joint Task Force One, 25 Sept. 1946.

Reingold, Nathan, ed. *The Sciences in the American Context: New Perspectives.* Washington: Smithsonian Institution Press, 1979.

Reitter, G. S., and Harrison S. Martland. "Leucopenic Anemia of the Regenerative Type Due to Exposure to Radium and Mesothorium." *American Journal of Roentgenology* 16 (1926):161–167.

Reynolds, Lawrence. "The History of the Use of the Roentgen Ray in Warfare." *American Journal of Roentgenology* 54 (1945):649–662.

Romer, Alfred. "Accident and Professor Röntgen." *American Journal of Physics* 27 (1959):275–277.

————. *The Discovery of Radioactivity and Transmutation.* New York: Dover, 1964.

————. *Radioactivity and the Discovery of Isotopes.* New York: Dover, 1970.

————. *The Restless Atom.* Garden City, N.Y.: Anchor Books, 1960.

Rose, Lisle A. *Dubious Victory: The United States and the End of the World War II.* Kent, O.: Kent State University Press, 1973.

Rowland, R. E., A. F. Stehney, A. M. Brues, M. S. Littman, A. T. Keane,

B. C. Patten, and M. M. Shanahan. "Current Status of the Study of ^{226}Ra and ^{228}Ra in Humans at the Center for Human Radiobiology." *Health Physics* 35 (1978):159–166.

Russ, Sidney. "A Personal Retrospect." *British Journal of Radiology* 26 (1953):554–555.

Russ, Sidney, et al. "The Injurious Effects Caused by X-Rays." *Journal of the Roentgen Society* 12 (1916):38–56.

Russell, E. R., and J. J. Nickson. "Distribution and Excretion of Plutonium." In *Industrial Medicine on the Plutonium Project: Survey and Collected Papers*, ed. Robert S. Stone. National Nuclear Energy Series. Div. IV, vol. 20. New York: McGraw-Hill, 1951. Pp. 256–263.

Sacher, G. A., Leon O. Jacobson, Edna K. Marks, and S. L. Tylor. "Biometric Investigation of Blood Constituents and Characteristics in a Population of Project Workers." In *Industrial Medicine on the Plutonium Project: Survey and Collected Papers*, ed. Robert S. Stone. National Nuclear Energy Series. Div. IV, vol. 20. New York: McGraw-Hill, 1951. Pp. 425–455.

Sarton, George. "The Discovery of X-Rays." *Isis* 26 (1937):349–364.

Schales, Fritz. "Brief History of ^{224}Ra Usage in Radiotherapy and Radiobiology." *Health Physics* 35 (1978):25–32.

Schull, William J., Masanori Otake, and James V. Neel. "Genetic Effects of the Atomic Bomb: A Reappraisal." *Science* 213 (1981):1220–1227.

Schultz, Vincent, and Alfred W. Klement, eds. *Radioecology*. New York: Reinhold; Washington: American Institute of Biological Sciences, 1961.

Scoville, Herbert, Jr. "The Atomic Bomb and the Resultant Phenomena." In *Radiological Defense*. Vol. 3: *A Series of Indoctrination Lectures on Atomic Explosion, with Medical Aspects*. Armed Forces Special Weapons Project. Washington, n.d. Pp. 16–22.

————. "Estimation of Hazard from Airborne Radioactive Material." Manhattan Engineer District, 17 Dec. 1945.

Seaborg, Glenn T. "Plutonium Revisited." In *Radiobiology of Plutonium*, ed. Betsy J. Stover and Webster S. S. Lee. Salt Lake City: J. W. Press, Department of Anatomy, University of Utah, 1972. Pp. 1–21.

Senecal, Vance E. "Du Pont and Chemical Engineering in the Twentieth Century." In *History of Chemical Engineering*, ed. William F. Furter. Washington: American Chemical Society, 1980. Pp. 283–301.

Serwer, Daniel Paul. "The Rise of Radiation Protection: Science, Medicine and Technology in Society, 1896–1935." Ph.D. dissertation. History of Science. Princeton University, 1977.

Sharpe, William D. "Chronic Radium Intoxication: Clinical and Autopsy Findings in Long-Term New Jersey Survivors." *Environmental Research* 8 (1974):243–383.

————. "The New Jersey Radium Dial Painters: A Classic in Occupational Carcinogenesis." *Bulletin of the History of Medicine* 52 (1979):560–570.

Sherwin, Martin J. *A World Destroyed: The Atomic Bomb and the Grand Alliance*. New York: Alfred A. Knopf, 1975.

Shonka, F. R., and H. L. Wyckoff. "Measurement of Nuclear Radiations." In

The Effects of Atomic Weapons, ed. Samuel Glasstone. Washington: Government Printing Office, 1950. Pp. 291–311.

Shurcliff, William A. *Bombs at Bikini: The Official Report of Operation Crossroads.* New York: Wm. H. Wise, 1947.

———. "Technical Report of Operation Crossroads." Report no. XRD-208. Joint Task Force One, 18 Nov. 1946.

Smith, Cyril Stanley. "Plutonium Metallurgy at Los Alamos during 1943–45." In *The Metal Plutonium*, ed. A. S. Coffinberry and W. N. Miner. Chicago: University of Chicago Press, 1961. Pp. 26–35.

Smith, R. Jeffrey. "Study of Atomic Veterans Fuels Controversy." *Science* 221 (1983):733–734.

Smyth, Henry DeWolf. *Atomic Energy for Military Purposes: The Official Report on the Development of the Atomic Bomb under the Auspices of the United States Government, 1940–1945.* Princton: Princeton University Press, 1945.

Spear, F. G. "The British X-Ray and Radium Protection Committee." *British Journal of Radiology* 26 (1953):553–554.

Stearns, J. C. "Counter or Defense Measures." Health Division. Report no. CH-259a. Supplement to CH-259. Metallurgical Laboratory, n.d.

Stone, Robert S. "General Introduction to Reports on Medicine, Health Physics, and Biology." In *Industrial Medicine on the Plutonium Project: Survey and Collected Papers*, ed. Robert S. Stone. National Nuclear Energy Series. Div. IV, vol. 20. New York: McGraw-Hill, 1951. Pp. 1–16.

———. "Health Protection Activities of the Plutonium Project." *American Philosophical Society Proceedings* 90 (1946):11–19.

Stone, Robert S., ed. *Industrial Medicine on the Plutonium Project: Survey and Collected Papers.* National Nuclear Energy Series, Div. IV, vol. 20. New York: McGraw-Hill, 1951.

Stone, Robert S., et al. "Report on Experimental Program of Health Division." Health Division, Metallurgical Laboratory, April 1943.

Stover, Betsy J., and Webster S. S. Lee, eds. *Radiobiology of Plutonium.* Salt Lake City: J. W. Press, Department of Anatomy, University of Utah, 1972.

Straume, T., and R. Lowry Dobson. "Implications of New Hiroshima and Nagasaki Dose Estimates: Cancer Risks and Neutron RBE." *Health Physics* 41 (1981):666–671.

Strope, Walmer E. "Investigation of Gamma Radiation Hazards Incident to an Underwater Atomic Explosion." Preliminary Design Section. Bureau of Ships, U.S. Navy, March 1948.

Szasz, Ferenc Morton. *The Day the Sun Rose Twice: The Story of the Trinity Site Nuclear Explosion, July 16, 1945.* Albuquerque: University of New Mexico Press, 1984.

Szilard, Leo. *Leo Szilard: His Version of the Facts. Selected Recollections and Correspondence.* ed. Spencer R. Weart and Gertrud Weiss Szilard. Cambridge: MIT Press, 1980.

Tannenbaum, Albert E., ed. *Toxicology of Uranium: Survey and Collected Papers.* National Nuclear Energy Series. Div. IV, vol. 23. New York: McGraw-Hill, 1951.

Tannenbaum, Albert E., and Herbert Silverstone. "Introduction and General Considerations." In *Toxicology of Uranium: Survey and Collected Papers*, ed. Albert E. Tannenbaum. National Nuclear Energy Series. Div. IV, vol. 23. New York: McGraw-Hill, 1951. Pp. 3–5.

————. "Summary of Experimental Studies: Relation to Uranium Poisoning in Man." In *Toxicology of Uranium: Survey and Collected Papers*, ed. Albert E. Tannenbaum. National Nuclear Energy Series. Div. IV, vol. 23. New York: McGraw-Hill, 1951. Pp. 45–50.

Tate, Merze, and Doris M. Hull. "Effects of Nuclear Explosions on Pacific Islanders." *Pacific Historical Review* 33 (1964):379–393.

Taylor, Lauriston S. "Brief History of the National Committee on Radiation Protection and Measurements (NCRP) Covering the Period 1929–1946." *Health Physics* 1 (1958):3–10.

————. "The Development of Radiation Protection Standards (1925–1940)." *Health Physics* 41 (1981):227–232.

————. "History of the International Commission on Radiological Units and Measurements (ICRU)." *Health Physics* 1 (1958):306–314.

————. "History of the International Committee on Radiological Protection (ICRP)." *Health Physics* 1 (1958):97–104.

————. *Organization for Radiation Protection: The Operations of the ICRP and NCRP, 1928–1974.* Report no. DOE/TIC-10124. U.S. Department of Energy and Department of Health, Education, and Welfare. Washington, 1979.

————. *Radiation Protection Standards.* Cleveland: CRC Press, 1971.

————. "Reminiscences about the Early Days of Organized Radiation Protection." In *Health Physics: A Backward Glance. Thirteen Original Papers on the History of Radiation Protection*, ed. Ronald L. Kathren and Paul L. Ziemer. New York: Pergamon Press, 1980. Pp. 109–122.

————. "Technical Accuracy in Historical Writing." *Health Physics* 40 (1981):595–599.

————. "X-Ray Protection." *Journal of the American Medical Association* 116 (1941):136–140.

Terry, Richard D. "Historical Development of Commercial Health Physics Instrumentation." In *Health Physics: A Backward Glance. Thirteen Original Papers on the History of Radiation Protection*, ed. Ronald L. Kathren and Paul L. Ziemer. New York: Pergamon Press, 1980. Pp. 159–165.

Tibbets, Paul W., with Clair Stebbins and Harry Franken. *The Tibbets Story.* New York: Stein & Day, 1978.

Trenn, Thaddeus J. "Rutherford on the Alpha-Beta-Gamma Classification of Radioactive Rays." *Isis* 67 (1976):61–75.

Truslow, Edith C., and Ralph Carlisle Smith. "Beyond Trinity." In *Project Y: The Los Alamos Story*. The History of Modern Physics, 1800–1950. Vol. 2. Los Angeles and San Francisco: Tomash, 1983. Part II. Published version of following report.

————. "Manhattan District History, Project Y, the Los Alamos Project." Vol. 2: "August 1945 through December 1946." Report no. LAMS-2532 (Vol. II). Los Alamos Scientific Laboratory, completed 1947, issued 1961.

Tsuzuki, Masao. "Experimental Studies on the Biological Action of Hard Roentgen Rays." *American Journal of Roentgenology* 16 (1926):134–150.

Tullis, John C. "The Gross Autopsy Findings and a Statistical Study of the Mortality in the Animals Exposed at Bikini." Appendix no. 3 to the Final Report. Naval Medical Research Section, Director of Ship Material, Joint Task Force One. 1 March 1947.

Tullis, John C., and Shields Warren. "Gross Autopsy Observation in the Animals Exposed at Bikini: A Preliminary Report." *Journal of the American Medical Association* 134 (1947):1155–1158.

Turner, G. L'E. "Röntgen." In *Dictionary of Scientific Biography*, ed. Charles Coulston Gillispie. Vol. 11. New York: Charles Scribner's Sons, 1975.

Uhl, Michael, and Tod Ensign. *GI Guinea Pigs: How the Pentagon Exposed Our Troops to Dangers More Deadly Than War: Agent Orange and Atomic Radiation.* Chicago: Playboy Press, 1980.

Underwood, E. Ashworth. "Wilhelm Conrad Röntgen (1845–1923) and the Early Development of Radiology." In *Sidelights on the History of Medicine*, ed. Zachary Cope. London: Butterworth, 1957. Pp. 223–241.

United States Atomic Energy Commission. *Atomic Energy and the Life Sciences.* Washington: Government Printing Office, 1949.

United States Strategic Bombing Survey. "The Effects of the Atomic Bombings of Hiroshima and Nagasaki." Washington, 19 June 1946.

———. *The Effects of the Atomic Bomb on Hiroshima.* 3 vols. Washington: Government Printing Office, 1947.

———. *The Effects of the Atomic Bomb on Nagasaki.* 3 vols. Washington: Government Printing Office, 1947.

———. "Summary Report (Pacific War)." Washington, 1 July 1946.

Vacirca, S. J. "Radiation Injuries before 1925." *Radiologic Technology* 39 (1968):347–352.

Voegtlin, Carl, and Harold C. Hodge, eds. *Pharmacology and Toxicology of Uranium Compounds, with a Section on the Pharmacology and Toxicology of Fluorine and Hydrogen Fluoride.* National Nuclear Energy Series. Div. VI, vol. 1, parts 1–4. New York: McGraw-Hill, 1949–1953.

Voelz, George L., Louis H. Hempelmann, Jr., J. N. P. Lawrence, and William D. Moss. "A 32-Year Follow-up of Manhattan Project Plutonium Workers." *Health Physics* 37 (1979):445–485.

Walsh, John. "A Manhattan Project Postscript." *Science* 212 (1981): 1369–1371.

Warren, Shields. "Hiroshima and Nagasaki Thirty Years After." *Proceedings of the American Philosophical Society* 121 (April 1977):97–99.

———. "The Pathological Effects of an Instantaneous Dose of Radiation." *Cancer Research* 6 (1946):449–453.

———. "The Physiological Effects of Radiant Energy." *Annual Review of Physiology* 7 (1945):61–74.

Warren, Shields, and R. H. Draeger. "The Pattern of Injuries Produced by the Atomic Bombs at Hiroshima and Nagasaki." *U.S. Naval Medical Bulletin* 46 (1946):1349–1353.

Warren, Stafford L. "The 1948 Radiological and Biological Survey of Areas in

New Mexico Affected by the First Atomic Bomb Test." Report no. UCLA-
32. Atomic Energy Project, University of California at Los Angeles, 17
Nov. 1949.

————. "The Role of Radiology in the Development of the Atomic Bomb." In
Radiology in World War II, ed. Kenneth D. A. Allen. Medical Department
in World War II: Clinical Series. Washington: Office of the Surgeon General, U.S. Army, 1966. Pp. 831–921.

Wasserman, Harvey, and Norman Solomon, with Robert Alvarez and Eleanor
Walters. *Killing Our Own: The Disaster of America's Experience with Atomic
Radiation.* New York: Dell, 1982.

Watts, Richard J. "Hazards of Trinity Experiment, Section II." Los Alamos Scientific Laboratory, 12 April 1945.

————. "Health Instrumentation and Crater Activity Decay for the July 16th
Nuclear Explosion." N.d. Appendix I in Joseph G. Hoffman, "Nuclear
Explosion 16 July 1945: Health Physics Report on Radioactive Contam-
ination throughout New Mexico Following the Nuclear Explosion, Part
A: Physics." Report no. LA-626. Los Alamos Scientific Laboratory, 20
Feb. 1947.

Weisgall, Jonathan M. "The Nuclear Nomads of Bikini." *Foreign Policy* no. 39
(Summer 1980):74–98.

Wigner, E. P. "Protection against Radiations; Protection against γ Rays." Health
Division. Report no. CH-137. Metallurgical Laboratory, ca. 20 June 1942.

Wilhelm, Harley A. "Development of Uranium Metal Production in America."
Journal of Chemical Education 37 (Feb. 1960):59–67.

Winant, Frank I., Jr. "Command Aspects of Radiological Defense." In *Radiological Defense. Vol. 3: A Series of Indoctrination Lectures on Atomic Explosion,
with Medical Aspects.* Armed Forces Special Weapons Project. Washington, n.d. Pp. 102–109.

Wintz, Hermann, assisted by Walther Rump. *Protective Measures against Dangers
Resulting from the Use of Radium, Roentgen and Ultra-Violet Rays.* Document
C.H. 1054. League of Nations Health Organization. Geneva, 1931.

Wirth, John E. "Management and Treatment of Exposed Personnel." In *Industrial Medicine on the Plutonium Project: Survey and Collected Papers*, ed. Robert S. Stone. National Nuclear Energy Series. Div. IV, vol. 20. New York:
McGraw-Hill, 1951. Pp. 264–275.

————. "Medical Services of the Plutonium Project." In *Industrial Medicine on
the Plutonium Project: Survey and Collected Papers*, ed. Robert S. Stone.
National Nuclear Energy Series. Div. IV, vol. 20. New York: McGraw-
Hill, 1951. Pp. 19–35.

Wyden, Peter. *Day One: Before Hiroshima and After.* New York: Simon & Schuster,
1984.

Yavenditti, Michael J. "The American People and the Use of Atomic Bombs on
Japan: The 1940s." *Historian* 36 (1974):224–247.

Zirkle, Raymond E., ed. *Biological Effects of External X and Gamma Radiation.*
National Nuclear Energy Series. Div. IV, vol. 22B. New York: McGraw-
Hill, 1954.

INDEX